Practical
Methods in Ecology

D0565743

Practical Methods in Ecology

P. A. Henderson

Director
Pisces Conservation Ltd
Lymington
Hampshire
UK

and

Senior Research Associate
Department of Zoology
University of Oxford
Oxford
UK

Blackwell
Publishing

BLACKWELL PUBLISHING
350 Main Street, Malden, MA 02148-5020, USA
9600 Garsington Road, Oxford OX4 2DQ, UK
550 Swanston Street, Carlton, Victoria 3053, Australia

First published 2003

4 2006

Library of Congress Cataloging-in-Publication Data

Henderson,P.A.
Practical methods in ecology/P.A. Henderson.
p. cm.
Includes bibliographical references (p.) and index.
ISBN 1-4051-0244-6 (pbk.: alk. paper)
1. Ecology—Methodology. I. Title.
QH541.28 H46 2003
577′.028—dc21 2002026255

ISBN-13: 978-1-4051-0244-5 (pbk.: alk. paper)

A catalogue record for this title is available from the British Library.

Set in 10/12pt Plantin
by SNP Best-set Typesetter Ltd, Hong Kong
Printed and bound in Singapore
by Markono Print Media Pte Ltd

The publisher's policy is to use permanent paper from mills that operate a sustainable forestry policy, and which has been manufactured from pulp processed using acid-free and elementary chlorine-free practices. Furthermore, the publisher ensures that the text paper and cover board used have met acceptable environmental accreditation standards.

For further information on
Blackwell Publishing, visit our website:
www.blackwellpublishing.com

Contents

Preface

This book is intended for use by undergraduate biologists and their lecturers. My aim has been to introduce a wide variety of ecological methods and analytical techniques that are appropriate for studies in open grassland, woodland, small freshwater habitats such as ponds and streams, and the seashore. These habitats are probably those that are most accessible to students and most likely to be studied on field courses. I hope the book will be a source of ideas and advice for students who undertake an ecological project as part of their coursework. There are parts of the book that I hope will be useful for sixth form and high school students.

Ecology is a fascinating subject that uses the ancient skills of tracking, trapping, and fishing, the eighteenth- and nineteenth-century skills of classification and identification, and more recent computer-aided approaches of multivariate analysis and computer simulation. While each of us will develop skills in only some of these areas there are certain core ideas and skills that all ecologists require, and this book is aimed at these areas.

Practical Methods in Ecology is closely linked to *Ecological Methods*, the standard text written by Professor Sir Richard Southwood with whom I had the privilege and pleasure of working on the 3rd edition. *Ecological Methods* is probably too advanced for undergraduate teaching, although it is an invaluable reference. Both books have the same general organization of the subject, which was thought out by Dick Southwood while teaching ecology at Imperial College. Dick Southwood taught me as an undergraduate and the knowledge gained was a fine foundation from which to build a career as a professional ecologist. I hope, by using his insight into how to structure the subject, a little bit of his magic as a teacher will also be transmitted. Many of the subjects just touched upon in this book are covered in far more detail in *Ecological Methods*, which I hope will be made more accessible by the present work.

I would like to thank Dr Richard Seaby and Mr Robin Somes of Pisces Conservation Ltd for their help, particularly with the scientific computing, and my wife Claire who proof read the manuscript.

Peter Henderson

Chapter One

Planning and preliminary considerations

If you do not carefully plan ecological fieldwork your effort will be wasted. However, some unfocused preliminary work can be useful to help to form your ideas. It is impossible to plan a study without background information on the natural history of the organism and the nature of the habitat. While much can be gained from reading, undertaking computer searches, and talking to others with experience, it is often essential to do some preliminary sampling and try out alternative techniques to assess their relative merits. During a pilot study a record should be kept of the cost of each part of the sampling routine, normally expressed in man-hours. If you have a clear appreciation of the cost, you will not attempt a study that is beyond your resources. Having completed preliminary studies, the objectives of your study should be clearly stated and a sampling program designed. Your objectives should be written down and you should be able to describe them verbally in clear and simple terms. If you are unable to do this, you probably have confused and muddled objectives and you certainly will not be able to convey your ideas to others.

Armed with the information obtained during the preliminary investigations and a clear objective, the full sampling plan can be created. It is important to remember that ecological studies rarely go according to plan as the natural world is variable and the weather will frequently disrupt sampling. Further, all sampling methods have a certain range of population density over which they are appropriate and may become unsuitable if the population should greatly decline or rise. Therefore build contingencies into the plan. While some studies focus on the ecology of a single species or even race (autecology), others investigate whole communities or some part of them (synecology). Both types of study require the collection of data about populations and require consideration of the same general principles described below.

The need for sampling

As it is rarely possible to count or collect all the individuals in a population, the size of the population and its attributes such as mean size and age must be estimated by sampling. Much of the planning of an ecological study is concerned with ensuring that the samples accurately reflect the population or community as a whole.

The scale of the study

The size of the area to be studied needs to be determined. This area must be of sufficient size to adequately reflect the true nature of the target population or community. However, if it is too large you will waste resources and may not be able to complete the study.

Whether the area of study should be a single habitat (e.g. field, pond, woodland, or particular rock pool) or selected representatives of the habitat type from a wider geographic area will depend on whether an intensive or an extensive study is planned. Extensive studies have a low intensity of sampling per unit area or through time. They are frequently used to provide information on distribution and abundance for conservation or management programs. Intensive studies involve the repeated observation of the population of an organism with the intention of producing accurate estimates of population parameters. An intensive study would be needed, for example, to produce a life-table (see page 108).

Safety

During the planning stage you should identify potential hazards and plan how to work safely. You will find it useful to refer to Nichols (1983). When working on or close to water, or in remote localities, it is important not to work alone. Pay particular attention to risks associated with adverse weather and exposure, becoming lost, and the effects of heat and excess sun. You may also need to take preservatives such as formalin or other chemicals into the field. If so, consider carefully how they will be transported and how harmful exposure such as splashing formalin into your eyes will be avoided and dealt with. If you plan to handle wild animals ensure you have the correct protective equipment to avoid bites and scratches. Finally, get acquainted with the features and risks of the study area. Are there biting insects that will make concentration and careful measurement at certain times of day impossible? Are there parasitic mites in the vegetation just waiting for a soft-skinned mammal to pass? If these aggravations and dangers are recognized you will be able to dress defensively, use repellents and work at the best times of day so that field work is a pleasure rather than an ordeal.

Care for the environment

As your plans are developed always consider the potential harm caused to the habitat and its inhabitants by your study and minimize general disturbance.

Keep the number of organisms handled, removed and killed to a minimum, and take care not to trample on plants that are easily damaged. If animals are to be returned, ensure that they are released close to their point of capture and in a manner that will give them a good chance of survival. Consider carefully if any of the proposed methods are cruel and likely to cause unnecessary distress. In particular, if animals are to be killed, plan carefully how this will be done. Finally, ensure that you know the local regulations concerning the protection of habitats and endangered species and always obtain the consent of licensing authorities, landowners, etc.

Taxonomy

Are there any taxonomic difficulties? It is essential that you ensure that you can identify the target species and maintain a consistent taxonomy. As a study progresses taxonomic ability frequently improves. It may be necessary to revisit early samples to reassess species identifications. When possible, you should retain samples. Studies have failed when it became apparent that two or more species had been confused during the early period of sampling. Problems can also occur if only some life stages can be identified.

In community studies, you may need to identify or discriminate between large numbers of forms. Will this be feasible? The appropriate degree of taxonomic discrimination must be decided upon. It may not be essential to identify to the species level. For example, a study of stream invertebrates for the effects of pollution can be based on the presence of families and the calculation of a quality index such as the BMWP score (see Chapter 10). In community studies it is common to identify different groups to different taxonomic levels. For example, in freshwater studies stoneflies and dragonflies may be identified to species level while the larvae of small flies are recorded at the family level. Such mixed taxonomic resolution is often appropriate and the only practical option. However, you may need to consider what effect this will have on the analysis of species richness or diversity.

Sample sorting and species identification are often the most labor-intensive parts of a study and it may be useful to process a trial sample during a pilot study to assess the effort required.

Recording, labeling, and note taking

It is essential that you keep good records of your observations while in the field. Observations may be entered onto a tape recorder, palm-sized computer, digital camera, or onto paper. The durability of pencil on paper is second to none and unlike all the other devices listed above, it will even survive total immersion in water and will not require batteries. Always use notebooks of good quality paper that are designed to withstand water. If you use any form of electronic device always extract the data immediately upon returning from the field, ensure the data is not corrupted, and make backups and hard copy as appropriate. As a backup, notebooks should be photocopied, scanned or digitally photographed. To protect valuable data it may be useful to transmit it by e-mail to another locality. Data protection is not the only reason why information should be transcribed as soon as possible; field notes are often rather poorly written and will be better understood while events are still fresh in your memory. Never return to the field without having secured your data from the previous trip.

When taking notes remember to date the page, record who is present, and other general information such as the weather. When using photography, you will need to record the details of each photograph. Digital cameras are invaluable aids for ecologists. They allow you to instantly check that a good photograph has been taken and frequently allow details of the image to be entered as text or even a voice record.

Samples must be securely labeled. Usually you should place a pencil-and-paper label inside the sample and label the outside with a permanent marker. Paper of the correct quality must be used, as some papers will disintegrate when wet. Remember that a standard permanent marker is water insoluble, but the label might be lost if the sample is preserved in alcohol which leaks out. When labeling, use a numbering system that will not become ambiguous if part of the label is lost. For example, avoid roman numerals such as ii and iii. It is frequently advisable to check the labeling (and preservative) soon after returning from the field.

Data security and processing

A sampling plan needs to consider the processes involved in data acquisition, organization, analysis, and presentation. Smooth and rapid progress along this chain is aided by the use of computers, but only if the data can be easily transferred between software. If different software products are used, then compatibility must be considered. Both software and hardware capability need to be considered for each stage of the study. During data acquisition, it is important to assess the data storage and processing requirements. Automatic data collection devices, such as digital temperature recorders, can collect prodigious amounts of data if set to record at short intervals. Remember that data has only been acquired when it has been processed into a usable form. Data processing and input rates are often different. Data that takes many weeks to enter into a computer is often analyzed and plotted in a few seconds. Portable computers and palm-sized input devices allow field observations to be immediately stored in digital form and this may offer a way of streamlining the data-input process. Great care needs to be taken to ensure that data is not subsequently lost.

Ecological data is frequently arranged in spreadsheet software such as EXCEL. These programs are particularly appropriate for data that are naturally arranged in a grid, such as the species recorded from a number of samples. In many cases, appropriate statistical tests and plots can be undertaken within the spreadsheet program. If the data will need to be exported to a more specialized program for statistical analysis, you should ensure during the planning stage that the spreadsheet program is able to export the data in a suitable form.

Effect of the time of year on sampling

You must ensure that your proposed study will be undertaken at a suitable time of year. It is often important to sample when numbers are high. For plants you may need to work when they are in flower so that they can be easily identified. For seasonal insects such as butterflies the adults may only be abundant for a relatively short period of time each year. For annual species the life stage or age group that is chosen for study will determine the sampling period. The life stage of choice may be chosen because it is particularly easy to sample or possibly because it has a particularly important ecological impact. For insect pests, the best stage for sampling may be that most closely correlated with the amount of damage. For aquatic organisms, timing is often determined by the reproductive cycle. For example, in temperate waters, benthic surveys (the study of organisms living on or in the seabed) carried out in autumn will show a population dominated by recent recruits. The same survey carried out in the spring or early summer will show the resident community that has survived both competition for space and the rigors of the winter. While the life history stage to be sampled will depend on the objectives, the stage must not be one whose numbers change greatly with time and it must be present for a period of sufficient extent to allow the survey to be completed. Finally, the easier the stage is to sample and count the better.

Extensive surveys can result in different areas holding populations that differ in their development. The timing of blooms and larval production in marine plankton, for example, can vary by 1 month over 1 degree of latitude. The timing of the flowering of plants also shows clear latitudinal and altitudinal gradients. These differences may be used to advantage by allowing the different habitats to be sampled in succession.

Effect of the time of day on sampling

When observing birds and mammals the time of day when observations are taken is clearly important. However, it is not always realized that this can also be the case for invertebrates. The diurnal rhythms of insects may cause them to move from one part of the habitat to another. Many grassland insects move up and down the vegetation at certain times of the day or night and aquatic insects may emerge at a particular time. During the day quite a proportion of active insects may be airborne. Similarly, plankton also show diurnal rhythms in their vertical distribution and may concentrate in surface waters at particular times of day or night. Fish that remain hidden by day become active and vulnerable to trapping at night. There is a marked periodicity of host-seeking behavior in many blood-sucking invertebrates, a fact with which many of us are familiar, as just after sunset is often the worst time for mosquito attack. It may be that sampling problems can be overcome, or additional information gained, if work is undertaken at night, dusk, or dawn, rather than during conventional working hours.

Types of population estimate

If populations are to be estimated you will need to decide in what form the population size will be expressed. Estimates of population size can be expressed as absolute number, population intensity, relative number, or an index. This list of possible measures is ordered from the most difficult to the easiest to achieve. The objectives of the study will determine if a simple index of abundance will suffice or if greater effort must be made.

Absolute population is defined as the number of organisms per unit area or volume. It is almost impossible to construct a budget or to study mortality factors without the conversion of population estimates to absolute figures. While some organisms and habitats allow absolute population estimates to be obtained at reasonable cost, others, such as insects living amongst boulders in fast-flowing streams, are impossible to estimate.

Population intensity is the number of organisms per unit of habitat, e.g. per leaf, per plant, or per host. When the level of the population is being related to habitat availability or plant or host damage, it is more meaningful than an absolute estimate. It is also valuable when comparing the densities of natural enemies and their prey. However, the number of habitat units per unit area is potentially variable and may need to be assessed. Remember that a high intensity population may reflect either high absolute numbers or a shortage of hosts or habitats.

Relative estimates give an abundance estimate that cannot be related to a unit area or habitat and will only allow comparisons between similarly collected samples. Such comparisons can be either between different localities or at one locality through time. For many organisms they are the only estimates that can be achieved at a realistic cost. They are especially useful in extensive work on species distributions, monitoring changes in species richness, environmental assessments, and recording patterns of animal activity. The methods employed are either the catch per unit effort type or various forms of trapping, in which the number of individuals caught depends on a number of factors besides population density. There is no hard and fast line between relative and absolute methods, for absolute methods of sampling are seldom 100% efficient and relative methods can sometimes be corrected in various ways to give density estimates. Relative methods are important in applied research such as fisheries or game management where most of the available information may be derived from fishing or hunting returns.

Population indices are produced when the organisms themselves are not counted, but their products (e.g. frass, webs, exuviae, tubes, nests, pollen, pellets, fur) or effects (plant damage, footprints) are recorded. Such methods are frequently used for the study of rare or shy mammals such as otters or cats that may be infrequently observed.

Defining the habitat unit

Organisms only use some of the available space. Insects, for example, may only be found in association with their food plant and plants frequently require a particular soil and aspect. It is therefore essential to identify where the study organisms are to be found and the unit of habitat that should be sampled. The habitat unit could, for example, be the fleece of a sheep, a bag of grain, a rock in a stream, a single tree, or areas of grassland or muddy seabed. In a study of lichens it might even be individual gravestones. The identification of the habitat unit will often require both a literature review and some preliminary nonquantitative sampling.

If the sampling unit is a habitat feature such as a shoot, tree, or stone, it should meet the following criteria:

1 All habitat units in the study area must have an equal chance of selection for sampling.

2 The number and size of the habitat units should not change over the course of the study or any changes should be easily measured. For example, if the sampling unit is a young shoot you should take care that the number of shoots remains constant or any changes in number can be counted.

3 The proportion of the population using the sample unit as a habitat must remain constant. Many species move between different habitats and thus the proportion of the population in a particular habitat can change through time.

4 The sampling unit must lend itself to conversion to unit areas.

5 The sampling unit must be easily delineated in the field.

6 The sampling unit should be of such a size as to provide a reasonable balance between the variance in the number of organisms present and the cost.

7 The sampling unit must not be too small in relation to the organism's size, as this will increase edge-effect errors.

8 The sampling unit for mobile animals should approximate to the average ambit of an individual.

Quadrat sampling

When quadrat or unit area sampling, a square sampling unit is usually used. While other shapes such as a circle may reduce bias from edge effects, if the total habitat is to be divided into numbered sampling units (for random number selection), then circular units are impractical because of the gaps. Clearly, the larger the sampling unit is in relation to the size of the organism, the proportionally smaller the boundary edge effect. Anticipate edge effects by using a convention, e.g. only individuals crossing the top and left-hand boundaries are counted.

As a general principle, a higher level of reproducibility is obtained (for the same cost) by taking many small units rather than a few large ones. The main disadvantage of sampling by small units is the number of zeros that may result at low densities; this may produce unwanted restrictions on your analysis, e.g. you cannot take the logarithm of zero. Small sampling units may also enable precision to be increased by distinguishing between favorable and unfavorable microhabitats.

Having defined the habitat unit for study it is also essential to obtain information on its spatial extent during any pilot studies. For example, a grab lowered from a boat collects an unseen unit of seabed. If the objective is to study the benthic fauna of sand substrates, then the distribution of these substrates needs to be known before the sampling plan can be formulated. Similarly, a study of the insects of oak trees can only be planned if you have an appreciation of the distribution of the trees. Different sampling methods may be needed for preliminary and final sampling to ensure that the habitat distribution is understood. When benthic sampling, a pilot study might use dredges to obtain a general idea of animal presence and substrate distribution followed by grab sampling to estimate density in the selected habitat.

Subdividing the habitat unit

Animals and plants do not use their entire habitat equally, so during preliminary studies you should consider if the habitat units should be subdivided for sampling. The habitat must be considered in relation to the needs of the organisms under study. In a study of woodland birds, for example, the various levels of the trees – upper, middle and lower canopy – would be considered as potentially different divisions. In herbage or grassland, if leaves or another small sampling unit is being taken, the upper and lower parts of the plants might be treated separately. Subdivisions may be orientated with respect to an environmental gradient. If none is apparent, it is often convenient to divide the area into regions of regular shape, but this may be neither possible nor desirable. Aspect can be important. In the northern hemisphere north-facing rocks in the intertidal zone offer the coolest, most shady habitat available, and thus tend to favor red algae and its associated fauna. Conversely, species towards the northern limit of their range may prefer south-facing slopes. In aquatic habitats, depth, substrate, orientation with respect to flow and the degree of exposure are important. When sampling plants for insects, the amount of subdivision that is necessary varies greatly between species. While some insects may be distributed randomly over the foliage, others may be aggregated at a certain level and this may change with the seasons.

A hierarchical design is one in which, for example, a number of plants are sampled from each of a number of plots from within a number of fields. If a certain number of samples is collected randomly within each level of subdivision, this is often termed **nested random sampling** and is analyzed by a **nested analysis of variance** (see Underwood (1997), Zar (1996) or Sokal and Rohlf (1995) for an introduction to these methods in biology). There may be two or three hierarchical levels, rarely more.

Statistical considerations

Underwood (1997) gives a detailed description of the use of analysis of variance in ecological experimentation, and standard statistical textbooks such as those by Zar (1996) and Sokal and Rohlf (1995) are invaluable.

The number of samples

Never undertake an ecological study without considering how many samples will be required to meet your objective. When estimating mean abundance, the sample number cannot be accurately determined, but a working approximation can be obtained using an estimate or reasonable guess of the mean and variance of the population. These estimates may be derived from preliminary sampling or from published studies. Within a homogeneous habitat, in which the organisms under study are normally distributed, the number of samples (n) required is approximated by:

$$n = \frac{t^2 s^2}{(D\bar{x})^2} \tag{1.1}$$

where $\bar{x}$ = the mean number of organisms per sample, s^2 = sample variance, D = the proportional precision of the mean (i.e. to obtain an estimate within ±5% of the true value of the mean, then $D = 0.05$), and t = the "Student's t" obtained from standard statistical tables, but it is approximately 2 for $n > 10$ at the 5% level.

As an example, consider a study to estimate the density of small bivalve molluscs on a beach. During a preliminary visit 10 core samples were taken, giving a mean and variance of 5 and 10 animals per core. If it is desired to estimate the mean to a precision of ±10% then the number of core samples required is estimated as:

$$n = 2^2 \times 10/(0.1 \times 5)^2 = 40/0.25 = 160$$

Note that in the above calculation t is approximated as 2, which is quite accurate enough for such an estimate.

The distributions of many organisms are clumped and the assumption that the animals are normally distributed may be inappropriate. Southwood and Henderson (2000) have described methods of estimation when this is the case. However, provided you only use the estimated sample number as a guide, the assumption of normality will not produce a highly misleading estimate.

Another type of sampling program concerns the measurement of the frequency of occurrence of a particular organism or event, for example the frequency of occurrence of galls on a leaf. If it was found in a preliminary survey that 25% of the leaves of oak trees bear galls, the probability is 0.25. The number of samples (N) is given by:

$$N = \frac{t^2 p(1-p)}{D^2} \tag{1.2}$$

where p = the probability of occurrence (i.e. 0.25 in the above example), and D = the proportional precision of the mean (i.e. to obtain an estimate within ±5% of the true value of the mean, then $D = 0.05$).

Worked examples of the calculation of sample number for a variety of sampling programs are given by Greenwood (1996).

The number of samples per habitat unit (e.g. tree, rock, field, or puddle)

There are two aspects to be considered, firstly whether and exactly how the habitat unit should be subdivided and secondly the number of samples within each unit that should be taken for maximum efficiency. If it is decided that the habitat should be subdivided you should be aware that if the distribution of the population throughout the habitat is biased towards certain subdivisions, but the samples are taken randomly, **systematic errors** arise. This can be overcome either by sampling so that the differential number of samples from each subdivision reproduces in the samples the gradient in the habitat, or by regarding each part separately and correcting at the end.

Ideally, the within-subdivision sample variability should be much less than that between subdivisions. To determine the optimum number of samples per habitat unit (e.g. oak tree) (n), the variance of within-unit samples (s_s^2) must be compared with the variance of the between-unit samples (s_p^2) and set against the cost of sampling within the same unit (c_s) or of moving to another unit and sampling within it (c_p):

$$n_s = \sqrt{\frac{s_s^2}{s_p^2} \times \frac{c_p}{c_s}} \tag{1.3}$$

If the interunit variance, s_p^2 is the major source of variance and unless the cost of moving from unit to unit (e.g. oak tree to oak tree) is very high, n will be in the order of 1 or less (which means 1 in practice).

Often a considerable saving in cost without loss of accuracy in the estimation of the population, but with loss of information on the sampling error, may be obtained by taking randomly a number of subsamples that are bulked before sorting and counting. This is especially true where the extraction process is complex, as with soil samples.

The pattern of sampling

Again, it is important to consider the object of the program carefully. If the aim is to obtain estimates of the mean density, then it is desirable to minimize variance. However, if the distribution pattern is of prime interest then there is no virtue in a small variance. In order to obtain an unbiased estimate of the population, the sampling data should be collected at **random**, that is so that every habitat unit in the universe has an equal chance of selection. In the simplest form – the **unrestricted random sample** – the samples are selected by the use of random numbers from the whole area (universe) being studied (random number tables are in many statistical works, may be generated using a computer program or, failing these sources, the last two figures in the columns of numbers in most telephone books provide a substitute). The position of the sample site is based on two random numbers giving the distances along two coordinates; the point of intersection is taken as the center or a specified corner of the sample. If the size of the sample is large compared with the total area then the area should be divided into plots that will be numbered and selected using a single random number. Such a method eliminates any personal choice, as human bias in selecting sampling sites may lead to large errors.

However, a random choice method is not an efficient way to minimize the variance, since the majority of the samples may turn out to come from one area of the field. The method of **stratified random sampling** is therefore to be preferred for most ecological work; here the area is divided up into a number of equal sized subdivisions or strata and one sample is randomly selected from each strata. Alternatively, if the strata are unequal in size, the number of units taken in each part is proportional to the size of the part; this is referred to as self-weighting. Such an approach maximizes the accuracy of the estimate of the population.

When it is apparent that there are systematic variations in the density of the study organism across the study area, **stratified sampling** should be used. In general the strata are chosen to minimize within-stratum variance. Individual strata need not form continuous patches within the study area. When the habitat is stratified, biological knowledge can often be used to eliminate strata in which few animals would be found. Such a restricted universe will give a greater level of precision for the calculation of a mean than an unrestricted and completely random sample with a wide variance.

The other approach is the **systematic sample**, taken at a fixed interval in space (or time). In general, such spatial data cannot be analyzed statistically; however, it is often the method of choice in studies on the distribution of species or communities.

Biologists often use methods for site selection for random sampling that are less precise than the use of random numbers, such as throwing a stick or quadrat and sampling where it lands. Such methods are not strictly random; their most serious objection is that they allow the intrusion of a personal bias; quite frequently marginal areas tend to be undersampled as we are not inclined to throw towards a boundary. Bias may intrude due to causes other than personal selection. For example, grains of wheat that contain the older larvae or pupae of the grain weevil, *Sitophilus granarius*, are lighter than uninfested grains and may rise to the surface where samples are normally gathered. Similarly, the distribution and behavior of parasitized or sick animals may be such that they are more vulnerable to capture or observation.

Accuracy, precision, and completeness

Accuracy measures how close an estimate is to the real value and **precision** measures the reproducibility of the estimate. A decision on the accuracy and precision of population estimates is taken by considering both the objectives of the study and the variability of the system under study. For example, many species of insect pest exhibit 10- or even 100-fold population change within a single season and recruitment in fish stocks can vary 10-fold between years. For such species an estimate of population density with a standard error of about 25% of the mean, enabling the detection of a doubling or halving of the population, is sufficiently accurate for stock assessment studies. For life-table studies, a higher level of accuracy, frequently set at 10%, will be necessary.

Conservationists often seek to build species inventories. Here completeness will replace accuracy as a measure of quality. While for large mammals or birds the aim might be to record all resident species, for high diversity groups such as beetles the objective may be set at only 5–10% of the total fauna. Community studies often aim to generate summarizing statistics such as measures of diversity or species richness that can be used to compare localities or changes through time. The accuracy and precision of these estimates must be carefully considered if changes are to be detected at the desired resolution. It is often worthwhile to undertake numerical simulations using a computer as these can give an indication of the effort required to achieve a set level of accuracy. Such studies often demonstrate that the available resources will not allow the desired accuracy to be achieved. You will then need to decide if a lower accuracy is acceptable or the study objectives should be modified.

However, there is always much to be gained from field observations so never allow pessimistic contemplation of statistical niceties and accuracy to inhibit data collection. Excessive concerns about accuracy can always be used by those who favor the warmth of the hearth to being in the field. It should always be borne in mind that the law of diminishing returns applies to the reduction of sampling errors. In the long run, more knowledge of the ecology of an animal or plant may be gained by studying other areas, by making other estimates or by taking further samples than by straining for a very high level of accuracy in each operation. The one activity that will reward great care and effort is the sorting and picking of organisms from samples – remember that it is inevitable that the actual species list and individual abundances will lie above the recorded values, as some will almost inevitably be missed. You must always ensure that this task is given sufficient time.

Errors and confidence limits

The statistical errors of estimated parameters can usually be calculated and are referred to as the fiducial limits (the estimate (x) being expressed as $x \pm y$ where $\pm y$ is the fiducial limits). These are sometimes incorrectly referred to as "confidence" limits, but the distinction between the two terms is in practice unimportant. The fiducial limits are calculated for a given probability level, normally 0.05, which means that there are only 5 chances in 100 that the range given by the fiducial limits does not include the true value If more samples are taken the limits will be narrower, but the estimate may not move closer to the actual value. Biologists are often concerned that some of the assumptions about sampling efficiency may be incorrect. It is sound practice to compare estimates obtained using methods that make **different** assumptions. If the estimates are of the same order of magnitude, then you can have much greater confidence that the result is not misleading. Laughlin (1976) has suggested that ecologists may be satisfied with a higher probability level (say 0.2) and thus narrower fiducial limits for estimates based on more than one independent method.

The objective of a study is often expressed in terms of a hypothesis capable of scientific analysis. When a statistical analysis is anticipated you will need to create a null hypothesis for testing. For example, in a study of anthill distribution the null hypothesis might be that the mean anthill density is the same in open areas and in scrubland areas. A type I error is said to occur when we reject this null hypothesis when it is true. For our example we will have concluded that the density of anthills was significantly different when it was not. The opposite situation when we accept a null hypothesis when it is wrong is called a type II error.

The normal distribution and transformations

The most important and commonly used of the theoretical distributions is the normal or Gaussian distribution. This has a probability curve that is a symmetric bell-shape. The probability density of a normal variable $P(x)$ is:

$$P(x)\frac{1}{\sqrt{2\pi\sigma}}\exp\left[-\frac{1}{2}\left(\frac{x-\mu}{\sigma}\right)^2\right] \tag{1.4}$$

where μ is the mean and σ the standard deviation.

This distribution is symmetric about the mean and the shape is determined by the standard deviation, σ. It is important to ecologists because most of the common statistical tests assume the data to be normally distributed. While many distributions obtained from observation, for example the heights of men, are approximately normal, the spatial distribution of individuals is seldom, if ever, normal. The distribution of a field population could approach normality only if the individuals were randomly scattered and the population was so dense or the size of sampling unit so large that considerable numbers were present in each sample. Therefore, in contrast to other distributions described below, the normal distribution is not of interest to ecologists as a means of describing dispersion. Its importance arises solely from the fact that for many statistical methods to be applied the frequency distribution must be normal. Although analysis of variance is quite robust to deviations from normality, data whose frequency distribution is considerably skewed and with the variance closely related to the mean cannot be analyzed without the risk of errors. It is sometimes possible to use a simple transformation to "normalize" the data.

Simple transformations

Ecological data is usually skewed because species distributions hold a few highly abundant species and a long tail of low abundance forms. Such distributions are usually transformed by taking logarithms or square roots. For example, if the square root transformation were applied to 9, 16, and 64 they would become 3, 4, and 8, and it will be observed that this tends to reduce the spread of the larger values. The interval between the second and third observations (16 and 64) is on the first scale nearly seven times that between the first and second observations; when transformed, the interval between the second and third observations is only four times that between the first and second. It is thus easy to visualize that a transformation of this type would tend to "push" the long tail of a skew distribution in, so that the curve becomes more symmetrically bell-shaped.

It must be stressed that transformation does however lead to difficulties, particularly in the consideration of the mean and other estimates (see below). It should not be undertaken routinely, but only when the conditions for statistical tests are grossly violated.

The distribution of individuals in natural populations is such that the variance is not independent of the mean. If the mean and variance of a series of samples are plotted, they tend to increase together. Where sampling and other errors are fairly large it will usually be found adequate to transform the data from a regularly distributed population by taking the square of each number, while for a slightly contagious (clumped or aggregated) population take the square root of each value, or for a distinctly contagious population take the logarithm.

In order to overcome difficulties with zero counts in log transformations a constant (normally 1) is generally added to the original count (x); this is expressed as "log (x + 1)". This transformation can produce serious distortions and should be avoided when possible. It is customary to transform percentages to angles (arcsin transformation). In some fields, such as multivariate analysis (see Chapter 12), extreme transformations such as the 4th root have been advocated. Such transformations should be avoided.

In summary, the use of transformations can lead to problems and should not be routinely undertaken. The biological interpretation of estimates based on transformed data is often difficult. If data have been transformed, the means of the untransformed and transformed values should be presented.

Chapter Two

Estimating the reliability of estimates and testing for significance

The use of quantitative ecological data to make generalizations or support a particular interpretation of events in the natural world almost invariably requires some statistical analysis. The two main features that other ecologists need to understand if they are to have confidence in your results are the degree of error or uncertainty in any estimates and the significance of any differences that you have observed. It is the simple statistical techniques to undertake these calculations that will be presented here.

Some error is inevitable and you should not try to hide or attempt to downplay the errors or uncertainties in your estimates. The confidence intervals of a population estimate are frequently ±100% of the estimate and may seem disappointingly large, but it is not a source of shame or a demonstration of failure. It is simply a statement of fact. Uncertainty enters all scientific estimation because of measurement error. However, in ecology sampling normally introduces far more uncertainty. It is easy to see how a small number of samples can easily lead to considerable error in an estimate. The average score from a six-sided dice after many rolls is $(1 + 2 + 3 + 4 + 5 + 6)/6 = 3.5$. However, after I rolled a dice four times in succession I obtained the numbers 6, 5, 6, 4 giving an average score of 5.25. This small sample gave an estimate that was 1.75 too large. But my scores were not extraordinary. In ecology we are often forced to have small numbers of samples and this in turn means that we must accept wide confidence limits on our estimates. A further source of error is the inadequacy of our sampling regimes. For example, the population may be highly clumped and the sampling strategy only really effective for a randomly distributed population. Another common source of error is that our sampling efficiency varies with the density of the population. You should not get the impression that the situation is hopeless and there is no point even trying to estimate ecological parameters. The natural world is never at rest and so all ecological parameters are constantly varying. Therefore there is no such thing as the population size, but there is a range over which the population varies and there is much we can understand by putting limits on the range of populations.

Once we accept that all estimates are subject to uncertainty we immediately understand the need for statistical arguments if we are to decide if the magnitudes of the mean density of two different populations are truly different. Similarly statistical arguments are required to decide if two variables are correlated or to help us to fit the best equation to a relationship between two variables.

The statistical literature is vast and there are many good statistics textbooks. For ecologists two widely respected general books are *Introductory Biostatistics* by Zar (1996) and *Biometry: the principles and practice of statistics in biological research* by Sokal and Rohlf (1995). *Statistical Ecology in Practice* by Waite (2000) is a more specialized text. For biologists with little mathematical training *Practical Statistics for Field Biology* by Fowler and Cohen (1990) is well written and popular with students.

Almost all statistical calculations are now accomplished using either a computer program or electronic calculator and it is rare to calculate test statistics by hand. However you should always consider if the result seems sensible – it is easy to input the data wrongly and then blindly accept the output.

A simple set of data on the abundance of a common plant will be used to illustrate the methods presented below. The objective of the study was to estimate the density of the early purple orchid in three habitats and to determine if the density varied between sites. A 1 m^2 quadrat was used to estimate the density of early purple orchids in grazed grassland. A 100 m^2 area of habitat was delimited and the number of orchids within 10 randomly placed quadrats were counted. The random placing of the quadrats within the area was achieved by the use of a random number table to define the upper left-hand corner of each quadrat. This sampling protocol was repeated at three sites chosen because site 1 was heavily grazed by horses resulting in a very short lawn, site 2 was close to a footpath and badly trampled by pedestrians, and site 3 was close to a busy road and possibly affected by road pollution including salt applied to the road during the winter. The results obtained are shown in Table 2.1.

Descriptive statistics for a site

Table 2.1 presents the data obtained on the early purple orchid. At site 1 it can be seen that the mean or average density is 3.9 individuals per m^2. An inspection of the raw data shows that there was considerable between-quadrat variability. We wish to summarize this variability and give an estimate of the confidence limits for the estimated mean density.

A simple measure of the variation in the observed density is the range, which is simply the difference between the largest and smallest observation. For example at site 1 the range is $8 - 1 = 7$ individuals per m^2. This can be a useful measure, particularly when the observed distribution is far from random or normally distributed. This is the case for site 3 where the observed range of 43 shows that the plant distribution is very different from that at the other two localities.

The standard deviation is a common measure of the variability about the mean. This measure is particularly appropriate if the densities are normally

Table 2.1 Early purple orchid in 1 m^2 quadrats at three localities in the New Forest in April 2002: statistical results.

Statistic	Site 1 Heavily grazed by ponies	Site 2 Close to a heavily used footpath	Site 3 Close to a busy road junction
Quadrat no.			
1	4	0	11
2	8	0	8
3	1	1	7
4	3	0	6
5	1	1	47
6	1	0	11
7	5	1	4
8	4	0	15
9	5	1	38
10	7	3	16
Mean density (no. per m^2)	3.9	0.7	16.3
Range (no.)	7	3	43
Standard deviation (no. per m^2)	2.47	0.95	14.47
Standard error (no. per m^2)	0.78	0.3	4.58
Median (no.)	4	1	5
10th percentile (no.)	1	0	5
25th percentile (no.)	1	0	7
75th percentile (no.)	5	1	16
90th percentile (no.)	7.5	2	42

distributed, as the standard deviation can be used to calculate the probability of a deviation from the mean of more than a certain amount occurring by chance. For example, there is less than a 5% probability that a density of greater than 1.96 times the standard deviation away from the mean would be expected by chance. For site 1 this gives the prediction that a density of greater than $1.96 \times 2.47 + 3.9 = 8.74$ individuals per m^2 would only be expected in less than 5 quadrats in every 100. The standard deviation is not always a useful measure, e.g. at site 3 the same calculation, $1.96 \times 14.47 + 16.3 = 44.66$ individuals per m^2, would indicate that the density observed in 1 of only 10 quadrats (47) would be unlikely to be observed in such a small number of quadrats. The reason it was observed was that the plants were highly clumped and thus the density was not normally distributed. In fact, it is very rare for either plants or animals to be normally distributed, so statistical arguments based on the normal distribution are often best avoided.

The standard error of the mean is a measure of how closely the mean calculated from the sample of 10 quadrats approximates the true mean of the population. For sites 1 and 2 this is probably a useful measure, but as for the standard deviation the clumping at site 3 makes it a misleading measure of variability for this site.

Instead of using the mean and standard deviation we can summarize the density at the sites using the median and the 10th, 25th, 75th, and 90th percentiles (Table 2.1). The median is the "middle" observation, computed by ordering all observations from smallest to largest, dividing the sequence into two and selecting the largest value from the half holding the smallest numbers. For example at site 1 the quadrats in order of abundance give the sequence 1, 1, 1, 3, 4, 4, 5, 5, 7, 8 for which the median is 4 individuals per m^2. The percentiles are calculated in similar fashion so that the 75th percentile is the observed density in the quadrat that has a higher density than 75% of the quadrats sampled. These measures for summarizing abundance are usually the most suitable for ecological data. They are particularly appropriate when the densities between sites are compared using nonparametric tests, as described below.

The median, the percentiles and the range are often presented graphically as a box plot (Fig. 2.1). It can be seen that this diagram clearly shows that the median density and range of density observed at site 3 are both greater than at sites 1 and 2.

Comparing two sites

Parametric and nonparametric methods

Parametric tests assume samples were drawn from normally distributed populations and are based on estimates of the population means and standard

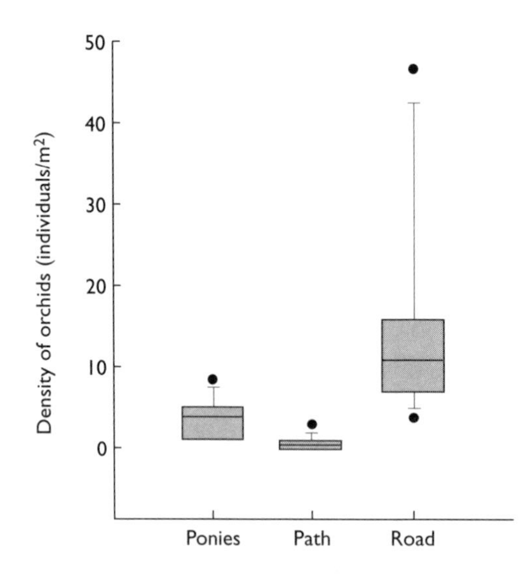

Figure 2.1 A box plot for the New Forest orchid data presented in Table 2.1. The median, percentiles and range of each site are plotted on the graph. The 25th and 75th percentiles for each site form the upper and lower limits of the filled boxes. Within the boxes the median is represented as a line. The vertical lines, where above zero, show the 10th and 90th percentile values. The maximum range in the data is shown by the black dots. The differences in the density of orchids observed at each site are clearly represented.

deviations. Nonparametric tests do not assume that the samples were drawn from a normal population. Instead, they perform a comparison based on ranked magnitude of each observation. Almost no data on the abundance or distribution of plants or animals will conform to a normal distribution, so nonparametric tests are most appropriate for ecological studies. It is generally harder to prove that two populations are significantly different using nonparametric methods, but this will have the benefit that you are less likely to incorrectly claim sites are different.

The Mann–Whitney rank sum or *U* test

This should be used to test if the medians of two groups are significantly different. The calculation is demonstrated by comparing the medians of site 1 and site 2 in Table 2.1. The null hypothesis is that there is no significant difference in the median density of orchids at sites 1 and 2.

1 Count the number of observations in each group (site) – for site 1 $n_1 = 10$ and for site 2 $n_2 = 10$. If the number of observations was different then n_1 would be the number of observations in the smaller group.

2 Rank the observations from both sites so that the smallest density is given a value 1 and tied values are given the same average rank, and sum the total rank for each group (see table below).

3 If R_1 is the summed rank for group 1 and R_2 is the summed rank for group 2 calculate U_1 and U_2 as follows:

$$U_1 = (n_1 \times n_2) + \left(\frac{n_1(n_1+1)}{2}\right) - R_1 \tag{2.1}$$

and

$$U_2 = (n_1 \times n_2) - U_1 \tag{2.2}$$

so for our example:

$$U_1 = (10 \times 10) + (10(10+1)/2) - 145.5 = 100 + 55 - 145.5 = 9.5$$

$$U_2 = (10 \times 10) - 9.5 = 90.5$$

Site 1 density	Ranks for site 1	Site 2 density	Ranks for site 2
4	15.5	0	3
8	20	0	3
1	9	1	9
3	13.5	0	3
1	9	1	9
1	9	0	3
5	17.5	1	9
4	15.5	0	3
5	17.5	1	9
7	19	3	13.5
Sum	**145.5**		**64.5**

4 The smaller of U_1 or U_2 is taken as the test statistic which is compared with the critical values obtained from statistical tables. The critical value for $n_1 = n_2 = 10$ is 23 which is larger than 9.5. It is therefore concluded that the median density of orchids is significantly different.

Comparing the medians of a number of samples – the Kruskal–Wallis test

This should be used to test if the medians of three or more groups are significantly different. Here we will compare the medians at all three sites given in Table 2.1:

1 Rank all the values as in the U test above. The lowest value is given a rank of 1 and tied values are given an average rank.
2 Sum the ranks for each sample, R_i.
3 Calculate the test statistic:

$$H = \frac{12}{N(N+1)} \sum_{i=1}^{m} \frac{R_i^2}{n_i} - 3(N+1) \tag{2.3}$$

where N is the total number of observations in each group and m is the number of groups (sites).

For the orchid data $N = 10 + 10 + 10 = 30$, $m = 3$, and the value of H (the test statistic) = 21.66. Critical values of H are obtained from tables of chi-squared values with $m - 1$ degrees of freedom. For the medians to be significantly different at the 5% level the critical value from chi-squared tables with 2 degrees of freedom is 5.99, indicating that the median density of orchids was significantly different between the three sites.

Measuring the correlation between variables – Spearman's rank correlation

We frequently wish to measure and test the significance of the correlation between two variables. Observations are correlated when a change in the value of one variable is reflected in a change in the value of the second. If an increase in one produces an increase in the second they are positively correlated. By contrast, if an increase in one results in a decrease in the second they are negatively correlated. The degree of correlation is measured by a correlation coefficient, which can range from −1 for a perfectly negative correlation via 0 for a situation where two variables are completely uncorrelated, to +1 for a perfect positive correlation. As an example, Table 2.2 presents data on the abundance of small (<5 cm long) trout and the depth of water in a small stream.

The working hypothesis is that small fish favor shallow water because large predatory fish are excluded. We therefore wish to find if there is a negative correlation between depth and trout density.

Table 2.2 Data on the distribution of trout with respect to depth in a small stream.

Average depth of the water (cm)	Rank for depth	Trout per m^2 (no.)	Rank for trout	Difference	Squared difference
10	1	14	7	–6	36
11	2	12	6	–4	16
18	4	6	4	0	0
16	3	8	5	–2	4
25	5	5	3	2	4
40	6	2	1	5	25
49	7	3	2	5	25
Sum (S)					**110**

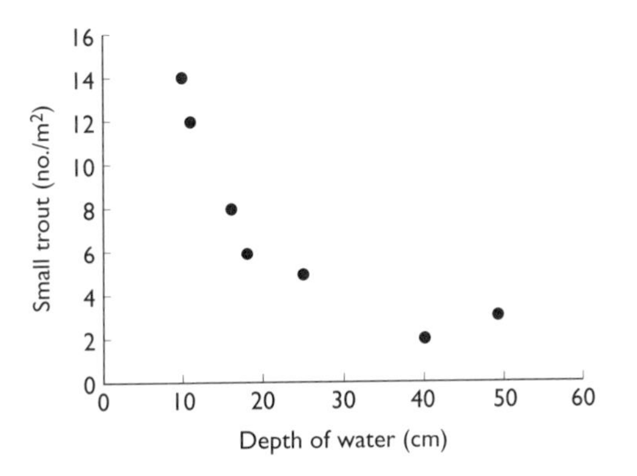

Figure 2.2 A simple scatter diagram of water depth against fish density. The graph shows the two variables to be negatively correlated. Take every opportunity to plot your data, as it will greatly increase your understanding.

A simple plot of water depth against trout density certainly suggests that the two variables are related (Fig. 2.2), as there is a tendency for more trout to be found in the shallower reaches. We wish to test if this negative correlation is significant. If the variables were normally distributed and linearly related we could use the Pearson product-moment correlation. However, these assumptions are unlikely to be true so the Spearman's rank correlation test will be used:

1 Rank each variable from smallest to largest giving the smallest a value of 1. Note that each variable is only ranked with itself and the numbers are not sorted into order (Table 2.2).

2 Calculate the difference between the ranked scores.

3 Calculate the square of the difference and find the sum.

4 Calculate the ranked correlation coefficient using:

$$R = 1 - \frac{6S}{n^3 - n} \qquad (2.4)$$

where n is the number of paired samples, which is 7 in our example, and S is the sum of the squared difference in the ranks.

For the trout example:

$$R = 1 - (6 \times 110/336) = 1 - (660/336) = -0.964$$

This value suggests a strong negative correlation between water depth and trout number. The significance can be found from tabulated critical values. If there are greater than 10 pairs of observations, critical values for Pearson's product-moment correlation can be used. In this case the negative correlation is highly significant.

Chapter Three

Sampling a unit of habitat – estimating absolute population number

This chapter presents techniques in which a piece of the habitat is sampled together with the organisms that are subsequently removed. These methods are often used to obtain estimates of species richness or diversity (see Chapter 10). If absolute population abundance is to be estimated, two separate measurements have to be made. We must estimate first the total number of organisms in the unit of the habitat sampled, and second the total number of these units in the whole habitat of the population under study.

It is important to remember that the errors in the population estimates obtained by these methods will normally lie below the true value. This is because extraction efficiency is less than 100%. Further errors can be large, as to obtain an estimate of total population you must multiply the mean count per sampling unit by the large number of units in the habitat.

Sampling discrete habitat units

Small microhabitats that are discrete units make ideal subjects for study. Good examples are flowers, fruits, seed heads, and cowpats. Flower heads of members of the Compositae such as the spear thistle, *Cirsium vulgare*, or the knapweed, *Centaurea nemoralis*, support small insect communities that are well known and can be used to study many features of community structure including competition and predator–prey relationships. Redfern (1968) describes in detail the insect fauna of spear thistle heads and how it can be studied. Similarly, the dung of cattle and other mammals presents an ideal (if somewhat unattractive) ecological unit within which to study interspecific relationships and the role of external factors on community structure (Skidmore 1985).

With such well-defined habitats it is possible to collect a set number of entire microhabitats from known localities and take them back to the laboratory for examination. However, you must consider the selection of the units to avoid bias in your sampling.

Such discrete habitats also lend themselves to video and photographic monitoring. For example, it may be possible to place a small video camera

Figure 3.1 A collapsible 50 cm^2 frame quadrat used for surveying short vegetation.

close to a flower head and record for extended periods insect activity and even predation.

Frame quadrats

One of the simplest and most widely used sampling devices is a square frame of wood or metal used to delimit an area within which the organisms are counted. Frame quadrats of 1 and 0.28 m^2 area are commonly used in vegetation surveys. A collapsible metal quadrat secured at the corners by butterfly bolts is illustrated in Figure 3.1. Wires stretched across the frame can further subdivide the area within the quadrat. The decisions that must be made on the size of the quadrat and where it should be placed in the habitat are discussed in Chapter 1.

When sampling plant communities it is often impossible to count the number of individuals of each species within a quadrat. A more frequently used measure is cover, defined as the percentage of the ground area occupied by the species within the quadrat. Community analysis programs such as TWINSPAN (see page 140) are designed to use percentage cover data. Even this measure can be very difficult to estimate and when large areas need to be surveyed

Figure 3.2 A point sampler used to estimate percentage cover in short vegetation.

abundance may be measured on the DAFOR scale (D, dominant; A, abundant; F, frequent; O, occasional; R, rare).

Point quadrats

Because the estimation of percentage cover using a frame quadrat is so subjective and prone to error, the point quadrat method was developed. The apparatus, which is illustrated in Figure 3.2, comprises a stand supporting a vertical bar drilled with holes at regular intervals through which 10 vertical needles can move. As the point of each needle is moved towards the ground a record is kept of the plant species that it touches. This procedure is repeated at a number of localities and the percentage cover calculated. For example, if 10 areas are sampled then there are $10 \times 10 = 100$ points for which data have been recorded. The percentage cover for a particular species is then given by:

$$\text{Percentage cover} = \frac{\text{Number of points that hit the species}}{\text{Total number of points observed}} \times 100$$

Using corers and high-sided quadrats

These techniques are suited to the sampling of soil, intertidal sand and mud, and freshwater soft sediments in shallow water. They can also be used in shallow, muddy bottomed ponds.

Core sampling a sandy or muddy beach

Animal abundance and diversity in intertidal sand or mud (soft) substrates can be surprisingly high. Even in polluted environments were diversity is low, abundance can reach extraordinary levels. For example, tubificid worm abundance in polluted parts of the tidal River Thames can reach 500,000 per m^2. A simple corer can be made from a 20 cm length of plastic drainpipe of about 4.5 cm diameter. One end should be sharpened and it is often useful to make a plunger that can be used to push the sample out. When larger samples are required, or fallen leaves or other debris is to be sampled, a quadrat can be made from a metal box or tank with the top and bottom removed. The following example is a brief study undertaken as part of an environmental impact statement for a new marina. Note that both core and quadrat samples were taken. The larger quadrat samples allowed the estimation of cockle, *Cerastoderma edule*, density that could not be adequately sampled using small cores. Cockles are bivalves up to 2 cm across the shell. The aim was to estimate the density of the main groups of macrofauna present so that the value of the beach to wading birds could be assessed. It was the size of food taken by birds that determined the size range of organisms that were extracted and counted.

An example of a beach survey

Methods

Sampling in Southampton Water, England, was carried out at low water spring tide, when the greatest area of the intertidal zone would be exposed. Two transects were laid out across the intertidal flats (Fig. 3.3). The total length of each transect was 150 m. The first sample site on each was some 2 m in from the low water mark; further sites were selected at 25 m intervals towards the shore. At each site two core samples and one quadrat were taken. Cores were obtained using a 6.5 cm diameter corer to a depth of 12 cm – roughly to the depth of the anoxic layer, below which there is unlikely to be many animals. This gave a volume of approximately 400 cm^3 of material per core. Quadrat samples were taken with a stainless steel quadrat, 32 cm by 25 cm, by 12 cm deep, giving an area of 0.08 m^2 per sample. The large-volume quadrat samples were first sieved using a 1 cm mesh to remove all the silt and sandy fraction, and the large shellfish and worms counted and identified. The core samples were sieved using a 0.5 mm mesh to remove silt and fine sand, and then preserved in 4% formaldehyde solution with Rose Bengal

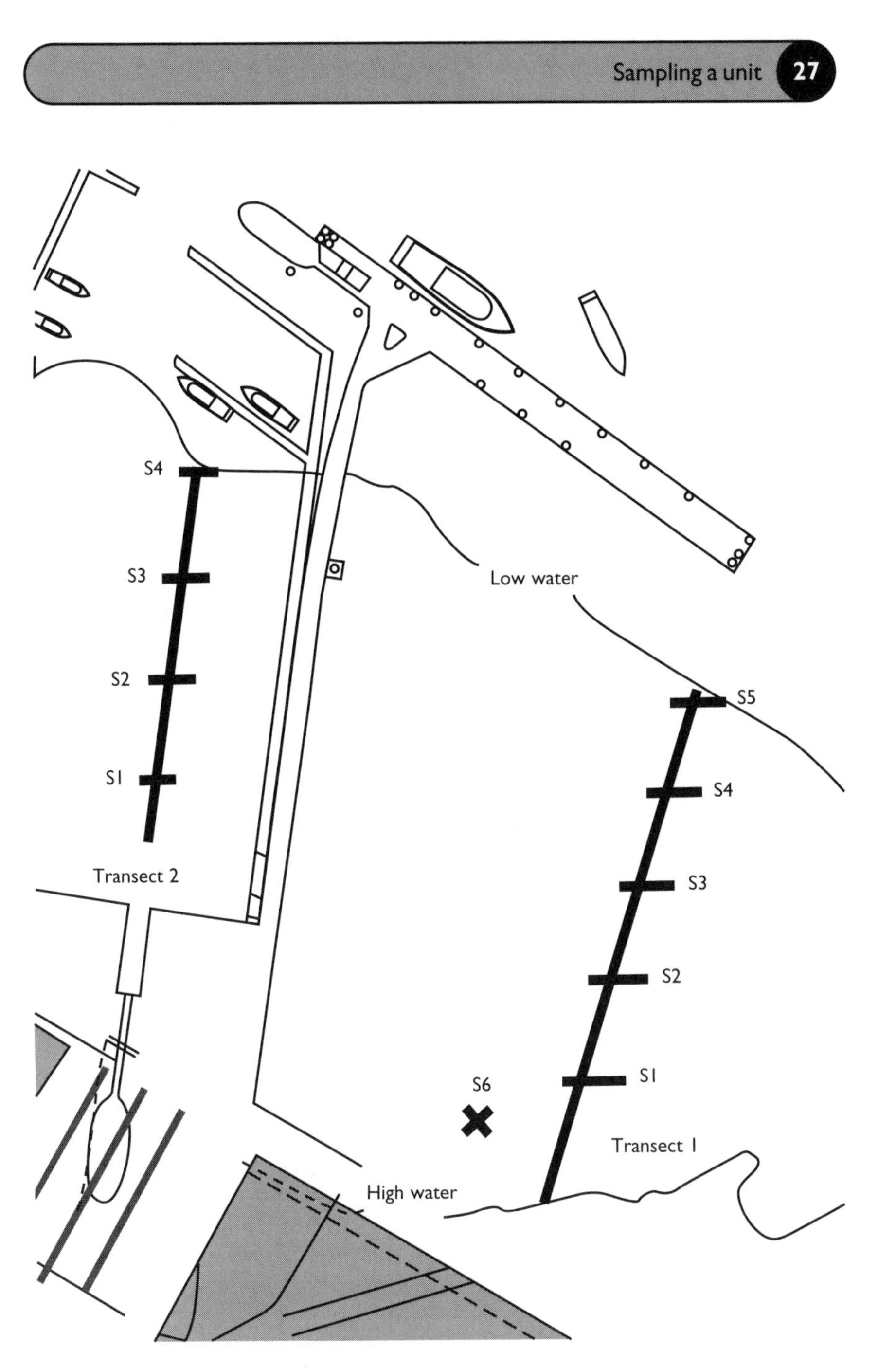

Figure 3.3 Map showing the position of the transects and the layout of the sampling stations for a beach survey in Southampton Water.

added to stain any organisms present. Subsequently, the core samples were sieved again, washed to remove the preservative, and the animals present picked by hand using soft forceps. The organisms were easily spotted as they were stained bright red.

Results

The species lists for the two littoral transects, with mean density as individuals per m^2, are presented in Tables 3.1 and 3.2. The former presents estimates of the density of the larger molluscs and crabs obtained by quadrat sampling and offers the most reliable estimates of cockle density. In Table 3.2 densities obtained by core sampling are presented, which are appropriate for the estimation of the density of worms and small crustaceans within the sediment. The replicate core samples gave similar results, suggesting that the substrate was adequately sampled to estimate density and record the majority of species present. Edible cockle, *Cerastoderma edule*, and the small snail, *Hydrobia ulvae*, dominated the fauna. Cockles were not evenly distributed within the study area. The area west of the existing jetty held the lowest density and the animals observed were at the lowest point on the beach (Table 3.1). East of the jetty, cockles were present at stations 1–5, although absent at station 6 which is closest to land.

While tubificid worms were the numerically most abundant group within the sediment, ragworm, *Hediste diversicolor*, were individually far larger and were probably the dominant infaunal species by biomass. Amphipod densities were low; *Corophium* species were only recorded from two stations. No rare or unexpected species were recorded during this survey. However, the amphipod, *Amphilochus neapolitanus*, which was found in low numbers, is an uncommon species, although easily overlooked. A small number of large American hard shell clams, *Mercenaria mercenaria*, was scattered on the lower levels of the beach. The littoral fauna can be characterized as a cockle–ragworm–*Hydrobia* community and is one of the dominant biotypes within Southampton Water.

The intertidal areas both west and east of the existing jetty (Fig. 3.3) held a similar fauna, although cockle densities west of the jetty (transect 2) were lower than in the eastern area. However, *H. ulvae* densities were considerably higher. *Hydrobia ulvae* is an extremely abundant species on intertidal mudflats and the densities recorded were not usual. The densities given in Tables 3.1 and 3.2 can be used as a rough guide to the densities of animals you should anticipate in productive waters. While the abundance of common animals on the littoral was typical for such sheltered beaches, the biodiversity found was low and the community was dominated by common and widely distributed species. Man has had a heavy impact on the study area. Much of the lower shore showed signs of recent bait digging for ragworm and the beach was scattered with considerable amounts of debris. A number of the species are foreign introductions including the hard shell clam, *M. mercenaria*, which is a native North American species. Southampton has long been an important

Table 3.1 Density of invertebrates from quadrat samples taken in the intertidal zone in Southampton Water.

		Density (individuals/m²)									Mean density (individuals/m²)		
Species	**Common name**	**T1 S1**	**T1 S2**	**T1 S3**	**T1 S4**	**T1 S5**	**T2 S1**	**T2 S2**	**T2 S3**	**T2 S4**	**T1**	**T2**	**All samples**
Cerastoderma edule	Edible cockle	26	22	17	17	13	—	—	6	5	237.5	34.375	147.22
Abra nitida	—	1	—	—	—	—	—	—	—	—	2.5	0	1.39
Abra alba	—	—	—	—	—	—	—	—	1	—	0	3.125	1.39
Littorina littorea	Edible winkle	—	—	—	7	—	—	—	—	—	17.5	0	9.72
Mercenaria mercenaria	Hard shell clam	—	—	—	—	2	1	—	—	—	5	3.125	4.17
Petricola holudiformis	American piddock	—	—	—	—	1	—	—	—	—	2.5	0	1.39
Carcinus maenas	Shore crab	—	1	1	1	—	—	—	—	2	7.5	6.25	6.94
Hediste diversicolor	Ragworm	P	P	P	—	—	P	—	P	—	—	—	—

T, transect; S, station; P, present but not quantified.

Table 3.2 The density of invertebrates from core samples taken in the intertidal zone in Southampton Water.

Group and species	Common name	Density (individuals/m²) T1 S1 C1	T1 S1 C2	T1 S2 C1	T1 S2 C2	T1 S3 C1	T1 S3 C2	T1 S4 C1	T1 S4 C2	T1 S5 C1	T1 S5 C2	T1 S6 C1	T1 S6 C2	T2 S1 C1	T2 S1 C2	T2 S2 C1	T2 S2 C2	T2 S3 C1	T2 S3 C2	T2 S4 C1	T2 S4 C2	Mean density (individuals/m²) T1	T2	All samples
Mollusca																								
Cerastoderma edule	Edible cockle	—	—	—	—	2	1	1	1	—	—	—	—	—	—	1	—	—	—	1	—	125.74	75.45	105.62
Abra nitida	—	1	—	—	—	—	—	—	—	—	—	—	—	—	—	—	—	—	1	—	—	25.15	37.72	30.18
Abra alba	—	—	—	—	—	—	—	—	—	—	—	—	—	—	—	—	—	2	1	1	—	0.00	150.89	60.36
Abra sp.	—	—	—	1	—	—	3	1	—	—	—	—	—	—	—	—	—	—	—	—	—	125.74	0.00	75.45
Littorina littorea	Edible winkle	—	—	—	—	1	—	—	—	—	—	—	—	—	2	—	—	5	3	1	—	25.15	414.95	181.07
Hydrobia ulvae	Mud snail	7	16	7	29	5	11	1	2	—	—	5	6	9	0	58	69	40	37	7	0	2,238.20	8,298.95	4,662.50
Crustacea																								
Carcinus maenas	Shore crab	—	—	—	—	—	—	—	—	—	—	—	—	—	—	—	—	—	1	—	1	0.00	75.45	30.18
Cyathura carinata	—	2	4	26	14	1	1	—	—	—	—	—	—	—	—	—	—	—	2	1	2	1,207.12	188.61	799.72
Microdentopterus sp.	—	—	—	1	—	1	—	—	—	—	—	—	—	—	—	—	—	—	—	—	—	50.30	0.00	30.18
Amphipoda indet.	—	—	—	2	1	—	—	—	—	—	—	—	—	—	—	—	—	—	—	—	—	75.45	0.00	45.27
Amphilochus cf. *neapolitanus*	—	—	—	2	1	1	—	1	—	—	2	—	—	—	—	1	—	—	—	—	—	176.04	37.72	120.71
Corophium volutator	—	—	—	—	—	—	—	—	—	—	—	—	3	—	—	—	—	—	—	—	—	75.45	0.00	45.27
Corophium sp.	—	—	—	—	—	—	—	1	—	—	—	—	—	—	—	—	—	—	—	—	—	25.15	0.00	15.09
Amphipoda indet.	—	—	—	—	—	—	—	—	—	1	—	—	—	—	—	1	—	—	—	—	—	25.15	37.72	30.18
Jaera albifrons grp	—	—	—	—	—	—	—	—	—	—	—	—	—	—	—	—	—	1	—	—	—	0.00	37.72	15.09
Annelida																								
Tubificoides	—	77	56	30	25	102	185	56	67	—	—	46	19	88	92	100	44	155	191	72	120	16,673.35	32,516.80	23,010.73
Hediste diversicolor	Ragworm	2	3	—	—	—	2	—	—	—	—	6	21	7	8	2	1	1	0	2	4	855.04	943.06	890.25
Nephtys hombergi	—	—	—	—	—	—	—	—	1	—	—	—	—	—	—	—	—	—	1	—	—	25.15	37.72	30.18
Sabellid indet.	—	—	—	10	—	—	1	—	—	—	—	—	—	—	—	—	—	—	—	—	—	276.63	0.00	165.98
Monayunkia aestuarina	—	—	—	—	1	—	—	—	—	—	—	—	—	—	—	—	—	—	—	—	—	25.15	0.00	15.09
Amphorete acutifrons	—	—	—	—	—	2	—	—	—	—	—	—	—	—	—	—	—	—	—	—	—	50.30	0.00	30.18
Cirriformia tenticulata	—	—	—	—	—	—	—	1	3	2	1	3	10	—	—	—	—	—	—	8	—	502.97	301.78	422.49
Polydora sp.	—	—	—	—	—	—	—	—	—	—	—	—	—	—	1	—	—	—	—	—	—	0.00	37.72	15.09
Anemone – small juveniles	—	—	—	1	—	1	3	—	—	—	—	—	—	—	—	—	—	—	1	—	—	125.74	37.72	90.53

C, core; S, station; T, transect.

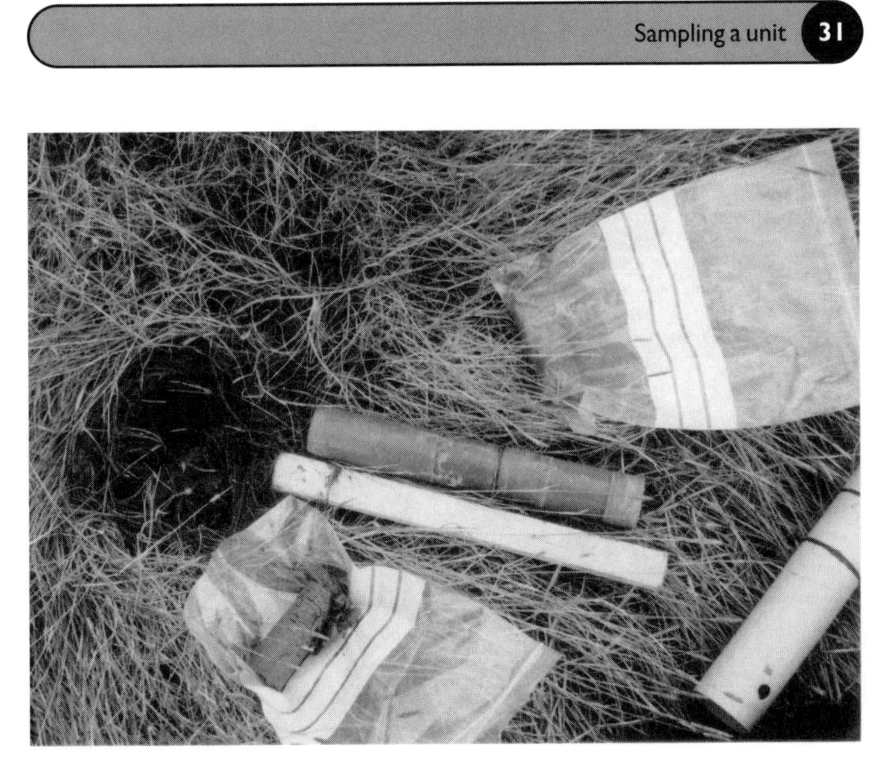

Figure 3.4 The apparatus required to take a soil sample: a section of plastic pipe sharpened at one end, a plunger to push the sample out, and self-seal plastic bags to hold the cores.

port and you should anticipate finding nonnative species in such areas in all parts of the world.

Core sampling soil or litter

Soil samples are usually taken with a corer; golf-hole borers or metal tubing sharpened at one end make simple corers. For small soil samples the sampler need be no more than a piece of plastic drainpipe with the edge at one end sharpened (Fig. 3.4). It has been suggested that some animals may be killed by compression when the core is forced from such "instruments", and furthermore it is highly desirable to keep the core undisturbed (especially for extraction by behavioral methods), so more elaborate corers have been developed. The depth to which it is necessary to sample varies with the animal and the condition of the soil; it will be particularly deep in areas with a marked dry season. With the O'Connor (1957) split corer (Fig. 3.5) the risk of compressing the sample by forcing it out of the corer is avoided. After the core has been taken the clamping band can be loosened, the two aluminum halves of the cover separated, and the sample exposed. The sample can then be easily divided into the different soil layers: litter, humus, upper 2 cm, etc. Fallen leaves and other debris are usually sampled with high-sided quadrats, such as a metal box with top and bottom missing and the lower edge sharpened. The

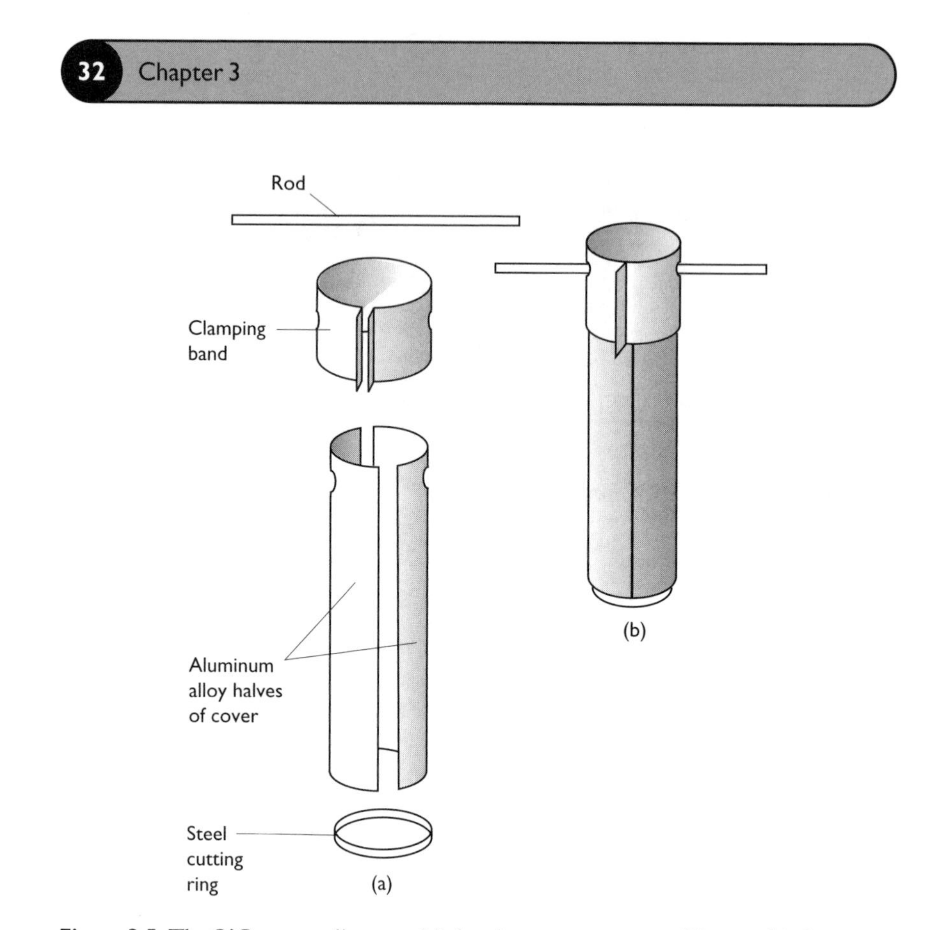

Figure 3.5 The O'Connor split corer: (a) showing compartments, (b) assembled.

quadrat is forced into the litter and the contents removed with a trowel or spade.

Once the samples have been obtained, the major problem for the ecologist is to extract the organisms of interest. Possible approaches include mechanical methods such as sieving (as was used above for intertidal invertebrates), flotation, elutriation, and dynamic or behavioral methods, all of which are discussed in detail in Southwood and Henderson (2000). The efficiency of these different approaches for the main arthropod groups found in soil have been assessed by Edwards (1991), and the recommended methods are presented in Table 3.3. Two standard and generally applicable methods for the extraction of arthropods from soil samples are Berlese–Tullgren funnels and Kempson bowl extractors.

The Berlese–Tullgren funnel is a simple to construct apparatus, and is drawn diagrammatically in Figure 3.6. It is used to extract large arthropods, e.g. Isopoda and Coleoptera, from bulky soil or litter samples. Most recent enhancements to the Berlese funnel have been designed to produce steep temperature and humidity gradients. Crossley and Blair (1991) describe a low cost design that can be used in a cool room with Christmas-tree lights used to produce a 20°C gradient.

Table 3.3 Recommended methods* for extraction of soil arthropods. (After Edwards 1991.)

	Soil type/system†								
	Peat			Clay/loam			Sand		
Group	W	P	A	W	P	A	W	P	A
Isopoda	AC	ABC	ABC	AC	ABC	ABC	All	All	All
Pauropoda	ABC	ABC	ABC	ABC	ABC	All	All	All	All
Symphyla	ABC	ABC	ABC	All	ABC	All	All	All	All
Diplopoda	ABC	ABC	ABC	ABC	ABC	All	All	All	All
Chilopoda	ABC	ABC	ABC	All	ABC	All	All	All	All
Acarina	ABC	ABC	ABC	ABCF	ABCDE	ACD	ABCF	ABC	ABC
Pseudoscorpionida	ABC	ABC	ABC	ABC	ABC	ABC	All	All	All
Araneae	ABC	ABC	ABC	ABC	ABC	ABC	ABC	ABC	All
Collembola	ABC	ABC	ABC	All	ABCD	ABCD	ABCD	ABCD	ABCD
Protura	ABC	ABC	ABC	ABC	All	All	All	All	All
Psocoptera	ABC	ABC	ABC	All	All	All	All	All	All
Thysanoptera	ABC	ABC	ABC	DE	DE	All	All	All	All
Hemiptera	ABC	ABC	ABC	All	All	All	All	All	All
Hymenoptera	ABC	ABC	ABC	All	All	All	All	All	All
Coleoptera	ABC	ABC	ABC	All	All	CDEF	CDEF	All	All
Diptera	ABC	ABC	ABC	All	All	All	All	CDEF	CDEF
Eggs/pupae	DEF	DEF	DEF	DEF	DEF	DEF	DEF	DEF	DEF

*All, all methods suitable; A, Tullgren funnel; B, Kempson extractor; C, air-conditioned funnel; D, Salt and Hollick flotation; E, other flotation; F, grease film extractor.
†W, woodland; P, pasture; A, arable.

The Kempson bowl extractor was developed by Kempson *et al.* (1963) and is suitable for the extraction of mites, Collembola, Isopoda and many other arthropods in woodland litter. Extraction rates of 90–100% are recorded for groups other than the larvae of holometabolous insects for which it is not efficient. The apparatus, which is drawn in Figure 3.7, consists of a box or shrouded chamber, containing an ordinary light bulb and an infrared lamp (250 W) that is switched on in pulses, at first only for a few seconds at a time and gradually increased until it is on for about a third of the time. The pulsing of the lamp is controlled by a time switch. The extraction bowls are sunk into the floor of the chamber. Each bowl contains a preservative fluid and is immersed in a cold water bath. A saturated aqueous solution of picric acid, with some wetting agent, was used by Kempson *et al.* (1963), but Sunderland *et al.* (1976) have found a 80 g/L solution of tri-sodium orthophosphate ($Na_3PO_4 12H_2O$) a cheap, safe, and suitable alternative. The litter is placed in a plastic tray above the bowl and is supported by two pieces of cotton fillet net laid on a coarse plastic grid. A fine black nylon screen covers the top of the tray. Kempson *et al.* (1963) give full instructions for the construction of this apparatus. A great advantage is the high humidity maintained on the lower surface of the sample. Extractions with this apparatus generally take about a week.

If the organisms differ in specific weight from the material in which they are

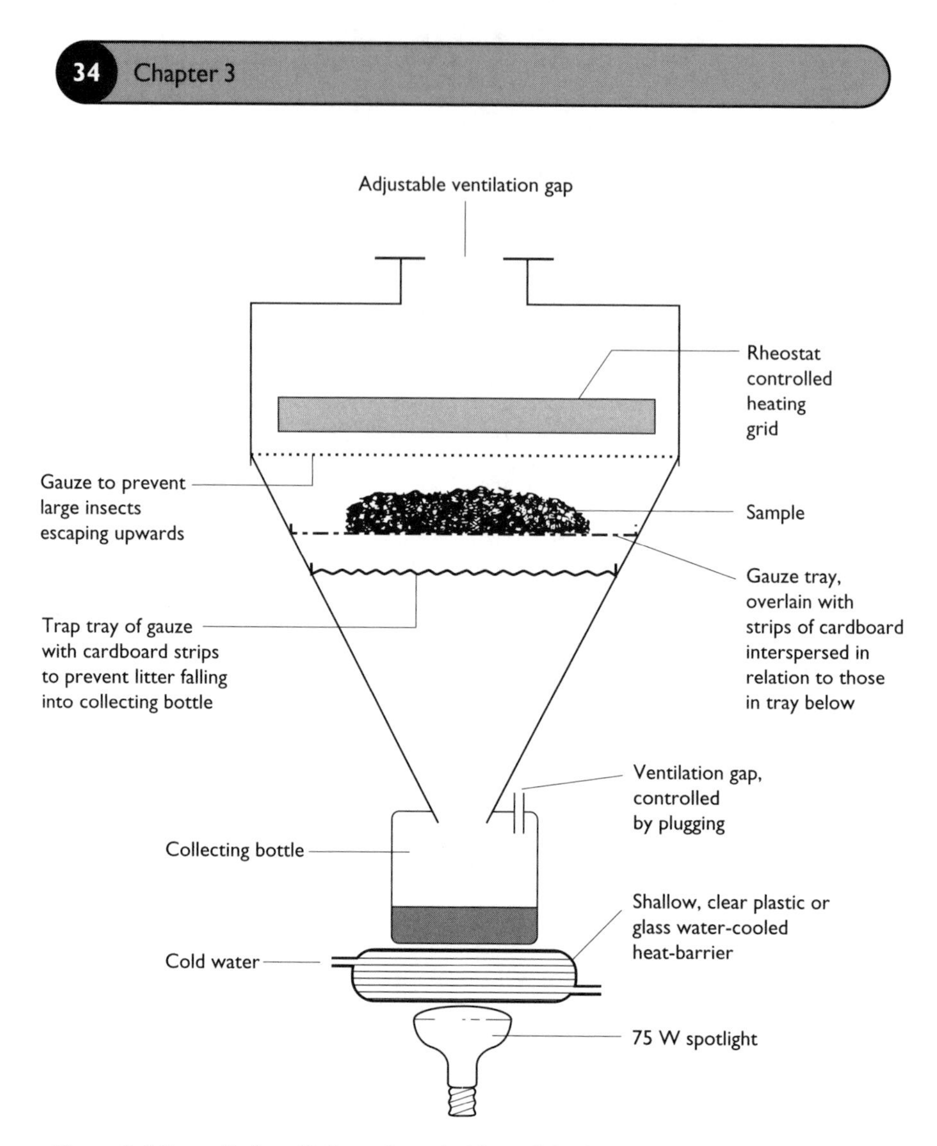

Figure 3.6 Large Berlese–Tullgren funnel with modifications.

dispersed then separation by flotation may be possible. The normal approach is to mix the sample into a liquid of specific gravity sufficient to allow the animals to float and the mineral particles to sink. Because organic debris also floats such methods are inappropriate for samples rich in organic detritus. A large variety of liquids have been used including magnesium sulphate, sodium chloride (salt), heptane, sugar solutions, and zinc chloride. For meiofaunal samples these solutions have been largely replaced by Ludox, a colloidal silica polymer. De Jonge and Bouwman (1977) extracted nematodes by mixing sediment samples with a 25% v/v solution of Ludox (specific gravity 1.39). The surface of the mixture was then covered with water to prevent gel formation and left for 16 hours, after which the surface layer was sieved and the organisms collected. Eleftheriou and Holme (1984) state that this method has also been used successfully for marine macrofauna. Sodium chloride

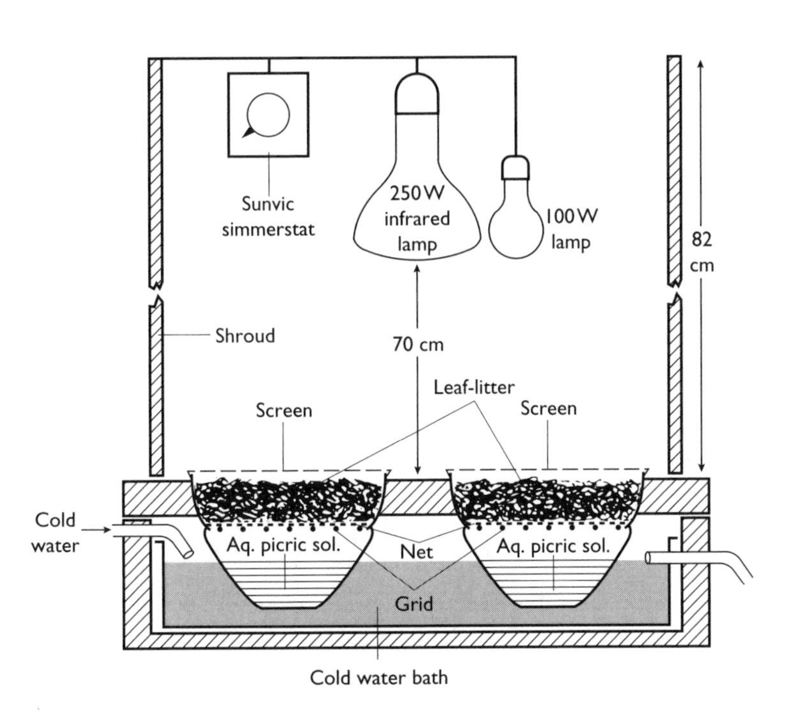

Figure 3.7 Kempson bowl extractor. (After Kempson *et al.* 1963.)

solution or brine has been widely used for arthropods and sucrose solution (specific gravity 1.12) has been found effective for insect samples from streams.

For semiarid soils, where behavioral extractors can be inefficient, Walter *et al.* (1987) recommend heptane flotation. The basic protocol is as follows (Walter *et al.* 1987; Kethley 1991).

1 Pour the sample into a stoppered measuring cylinder and add 1 liter of 50% ethyl alcohol and about 10 ml of heptane.
2 Replace stopper. Without shaking invert cylinder and allow the heptane to rise. Repeat this step twice.
3 Allow the cylinder to stand until the fine sediment has settled – a minimum 4 hours.
4 With minimal disturbance of sediment, decant the heptane layer into a sieve.
5 Rinse the sieve with 95% ethyl alcohol to remove the heptane and wash the sample into a sorting dish.

Grab sampling in lakes or the sea

Samplers with closing jaws are called grabs. They are often the method of choice for quantitative sampling of the fauna living within sand or mud. The

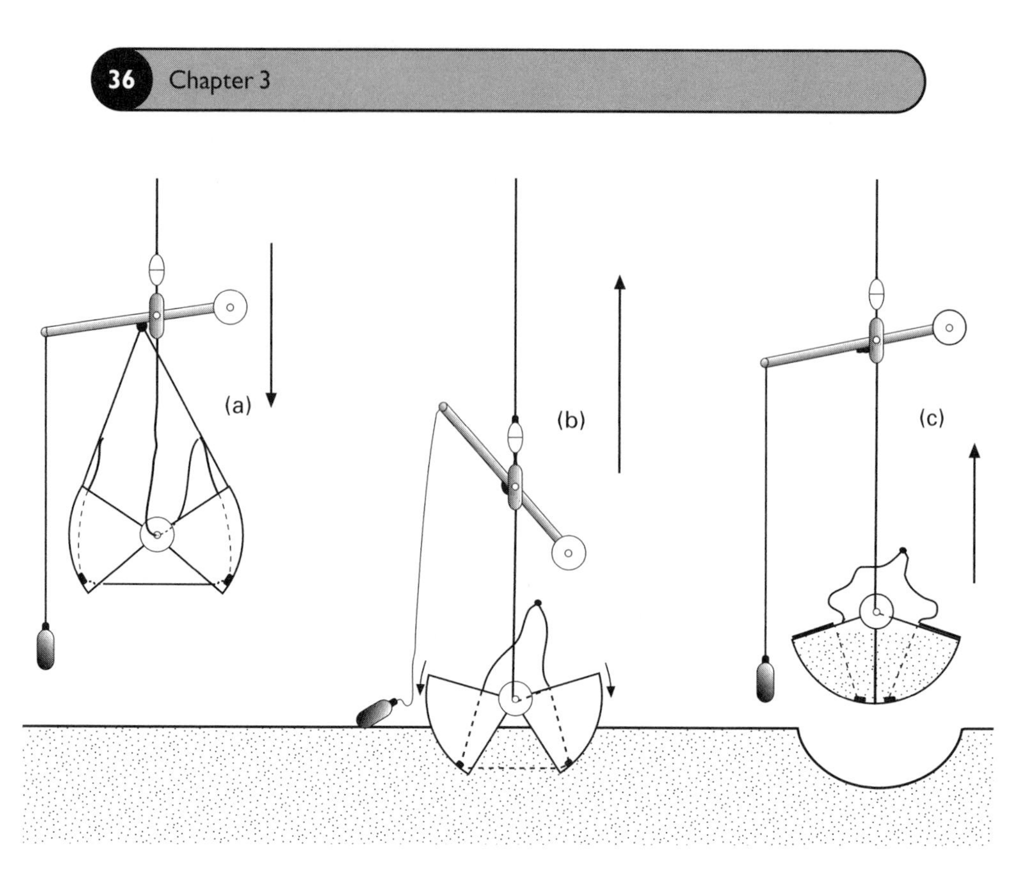

Figure 3.8 The operation of a Petersen grab (a–c) showing a method of closure that uses a counterpoised weight. (After Lisitsyn & Udintsev 1955.)

basic principle is to lower the grab to the substrate surface where the grab is activated so that the jaws cut into and enclose a sample of the substrate which can then be raised to the surface (Fig. 3.8). The fundamental problems in grab design are to ensure, firstly, that it rests correctly on the substrate so that when the jaws are closed the desired volume of substrate is taken and, secondly, that the jaws fully close to retain the sample during recovery. Stones and shells frequently stop the jaws from closing.

More than 60 types of grab have been described (Elliott & Tullet 1978). The most commonly used are variants of four basic designs: Petersen, van Veen, Birge–Ekman, and Smith–McIntyre (Fig. 3.9). The choice of grab will depend on depth, flow, substrate, weather (sea conditions), and the type of craft or platform from which it will be operated. Eleftheriou and Holme (1984) and Elliot and Tullet (1978) give extensive bibliographies of comparative sampling efficiency. The efficiency of capture of animals is only one criterion that needs to be considered. Others include substrate penetration characteristics, sample size, reliability of operation, and ease of handling.

In marine habitats with large boats and mechanical winches heavy grabs such as the Petersen (Petersen & Boysen Jensen 1911), Okean (Lisitsyn & Udintsev 1955), van Veen (van Veen 1933), Ponar (Powers & Robertson

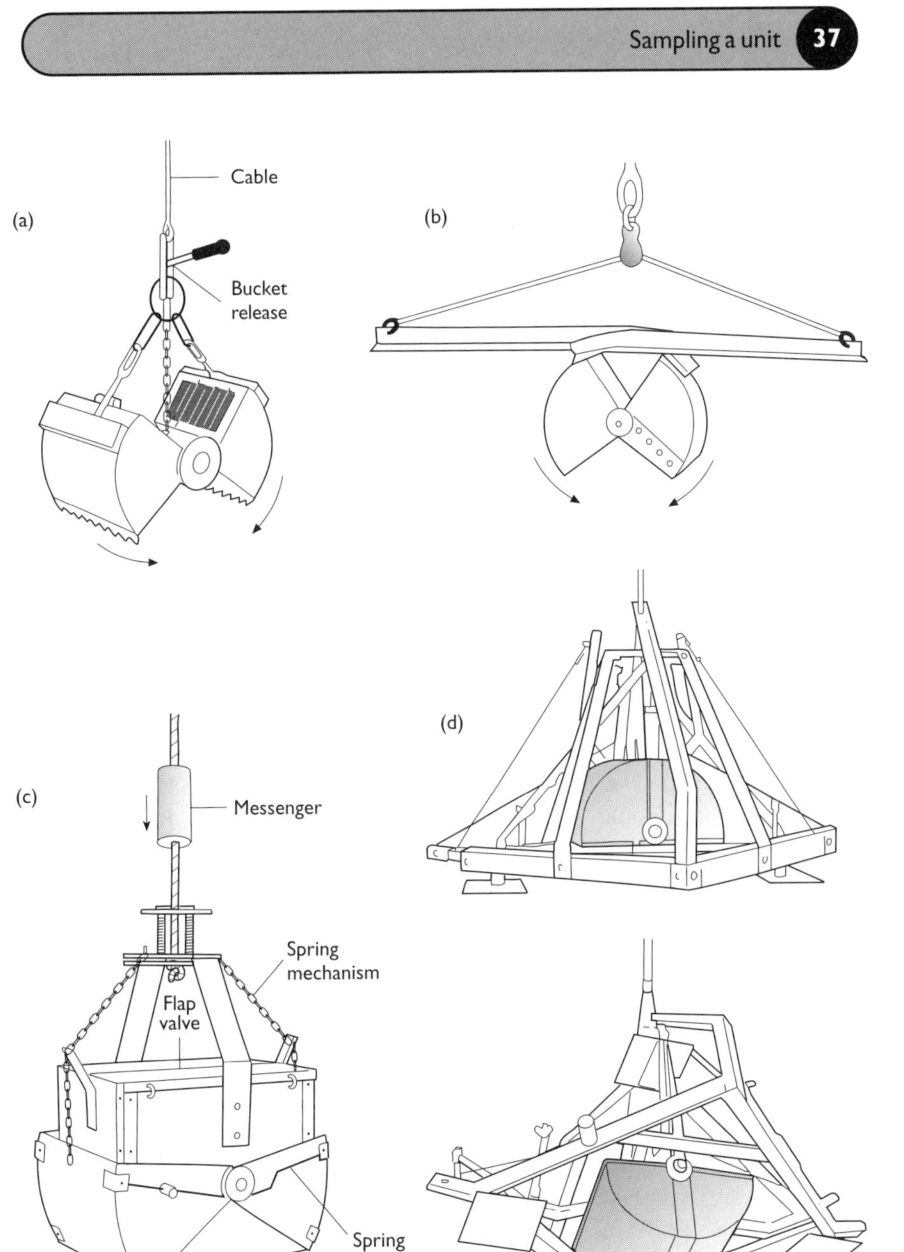

Figure 3.9 The basic types of grab design: (a) Petersen; (b) van Veen; (c) Birge–Ekman; (d) Smith–McIntyre.

Table 3.4 Efficiency of small grabs suitable for freshwater work measured as the percentage of pellets collected from a fine gravel substrate. (After Elliot & Drake 1981.)

Grab	Efficiency at the surface	Efficiency at c. 3 cm	Area sampled per sample (m^2)
Ponar	100	70	0.056
van Veen	87	56	0.1
Birge–Ekman	73	37	0.023
Allan	51	36	0.035
Friedinger	59	7	0.03
Dietz–Le Frond	22	26	0.016

1967), Hunter (Hunter & Simpson 1976), Smith–McIntyre (Smith & McIntyre 1954) or Day can be used. Grab sampling at sea requires experience and specialist guidance should be sought. A general introduction to the subject is given in Eleftheriou and Holme (1984). Marine grabs can rarely penetrate deeper than 10–15 cm and thus will not sample deep-burrowing animals. Further, they will rarely capture faster-moving epibenthic animals such as crabs, which may be quantified using photographic, video, or trawl methods.

Lighter grabs for use from small craft in ponds, rivers, and lakes include: van Veen (21 kg), Ponar (weighted 23 kg, unweighted 14 kg), Birge–Ekman (7 kg), Friedinger (8 kg), Dietz–Le Frond (La Fond) (21 kg), and Allan (8 kg). In water up to 3 m deep the Birge–Ekman and Allen grabs can be pole operated. Elliott and Drake (1981) compared the performance of all these grabs for freshwater work (Table 3.4). Ponar, Birge–Ekman, and Allen grabs performed well in muddy substrates (particle size 0.004–0.06 mm). They concluded that the Ponar grabs performed best with mud or fine gravel bottoms and were the only design to sample at >3 cm depth when small stones (8–16 mm largest dimension) were present. When stones (>16 mm largest dimension) were present only the weighted Ponar sampled to a depth of 2 cm. No grab performed well when the substrate was predominantly stones. The Ekman grab has been compared with various corers for sampling chironomid larvae in mud: Milbrink and Wiederholm (1973) found little difference, but Karlsson *et al.* (1976) found coring followed by separation by flotation recovered more larvae. (Chironomids are small flies with larvae that live in water.)

It is normal to coarse sieve grab samples in the field to reduce the volume of material that must be treated with preservative and transported to the laboratory for sorting. Leaves and stones can be inspected for attached organisms and if none are present discarded.

An example of a subtidal survey

Grab sampling was undertaken between 14 and 16 May 2001 in the Severn Estuary in the region between King Road and Flatholm Island (Fig.

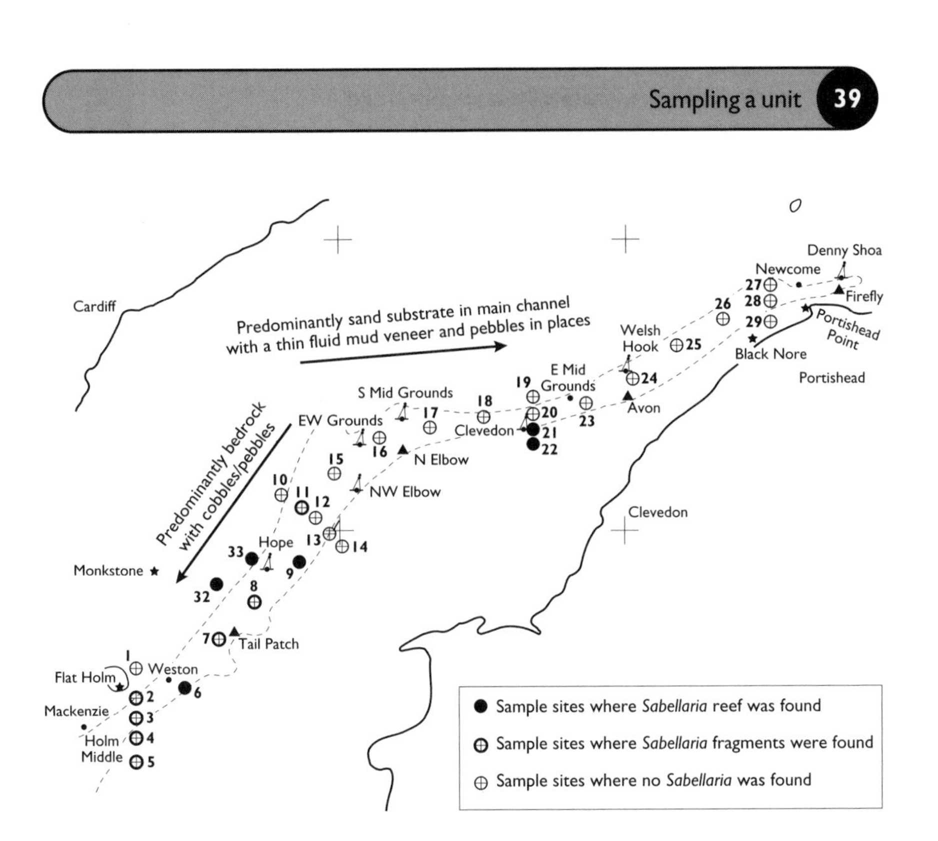

Figure 3.10 Map of the study area in the Severn Estuary showing the position of the sampling stations and the localities where evidence of *Sabellaria* reef was found.

3.10). Sampling transects of 3, 4, or 5 stations were set at 5 km intervals along the deep-water channel. Further individual sample stations were chosen at 1 km intervals between the main transects. In addition to the channel surveys, a small number of samples were collected at stations outside the main channel. Bad weather prevented the collection of these extra samples from the full length of the channel; however, it was possible to sample a number of sites on the Welsh (northwestern) side of the channel. Station position fixing was by a global positioning system (GPS). Figure 3.10 is a map of the region showing the position of the sampling stations.

Samples were collected using a 0.1 m^2 Day grab, which takes a semicylindrical bite from the seabed with an area of 300 by 330 mm. The penetration, and hence the volume of the sample, varies according to the composition and compactness of the substrate. Two grab samples were collected from each station. On being brought aboard, the grab was checked to ensure that it had deployed correctly; if this was not the case the sample was repeated. In instances where a stone had prevented the jaws of the grab from closing properly, grabbing was repeated until a valid sample was obtained. At certain stations, particularly at the western end of the channel, the grab repeatedly came up completely empty; when this occurred it was assumed that the substrate was bare rock.

Samples were inspected visually, and then emptied into a plastic tray, a photograph was taken and general features of the composition including the presence of mud, shell debris, organic matter, or pieces of *Sabellaria* reef were noted. Any *Sabellaria* pieces present were removed and preserved separately, as were any large stones. The presence of any encrusting organisms was recorded. The remaining sample was then washed through a 0.5 mm sieve and the retained material preserved by the addition of 40% formaldehyde solution with 1% Rose Bengal. Generally, the sand fraction of the sample would pass easily through the 0.5 mm sieve. However, towards the eastern end of the channel, the grain size was larger. This occasionally resulted in a sample not being sieved completely before the collection of the next sample was due. In these instances, the entire bulk of the sample was preserved for processing later.

The samples were quantitatively sorted, and all organisms extracted and preserved in formaldehyde. Any *Sabellaria* reef was dabbed dry and weighed, to provide an estimate of the quantity present. The animals in the samples were identified to species and counted. Where possible, actual numbers per sample were obtained, but it is not possible to quantitate colonial epifaunal organisms such as hydroids and bryozoans, and these were recorded as present or absent. It was also found impossible to identify to species the encrusting sponge (only one species found) or the anemones (two species found) from preserved samples. The number of *Sabellaria* worm individuals in the samples could not be fully quantified when large chunks of reef were present in the grab samples. However, the relative amount of *Sabellaria* reef present could be inferred from the weight of reef in each sample.

The results of the analysis were plotted on the map of the stations to show the localized distribution of *Sabellaria* in this region (Fig. 3.10).

Sampling invertebrates and small fish in ponds and streams

Sampling stream invertebrates by kick sampling

The majority of the animals in fast-flowing streams will be underneath the stones on the bottom. If a net is placed on the streambed against the flow and the substrate in front of the net agitated, any animals that enter the flow will be caught in the net. The simplest approach is to kick the stones. While kick sampling is frequently used in streams (Fig. 3.11), it will not yield an absolute estimate of population size as an unknown proportion of the invertebrate population will be caught. The Surber sampler comprises a net with an attached frame that delimits the area that will be agitated (Fig. 3.12). Comparison of the results from the Surber sampler with absolute counts of animals from buried trays showed that it caught only about a quarter of the population. Possibly, we should not consider it an absolute method, but it will give some idea of the numbers per unit area and no better method is known for the

Figure 3.11 Kick net sampling in a small English chalk stream. On a sunny day this is a pleasurable activity.

stony gravel beds of streams. Further, you should remember that there is no such thing as a perfectly efficient sampling method.

A typical application would be the comparison of invertebrate communities in different regions of a stream.

Hand net sampling of leaf litter in streams

Streams flowing through forest receive important inputs of leaves and other plant debris that may form banks on bends in the stream or behind obstructions. A great variety of insects, crustaceans and fish use this habitat, for which it offers both shelter and a feeding ground. It is a particularly difficult substrate to sample quantitatively and in shallow waters the most practical way to lift a portion of a substrate composed of forest litter may be with a hand net.

Figure 3.12 Three standard nets used for sampling freshwaters. The net with a long handle attached is a standard pond net. The conical net is a small plankton net suitable for lakes and ponds. The frame net or Surber sampler is used to delimit an area, which is kick net sampled; the displaced animals are washed downstream by the current into the net.

ea Example application

Henderson and Walker (1986) used shallow hand nets of 0.1 m^2 area to quantitatively sample the insect and fish infaunal community of Amazonian stream litter banks. The net was rapidly thrust into the side of a submerged litter bank and raised rapidly vertically to the surface. On the bank the leaves were carefully washed from the net and the fish and other large invertebrates present were identified and counted. The samples were collected from a regularly arranged grid of sampling stations allowing the distribution of the different species to be compared (Fig. 3.13).

(continued)

ea Example application (*continued*)

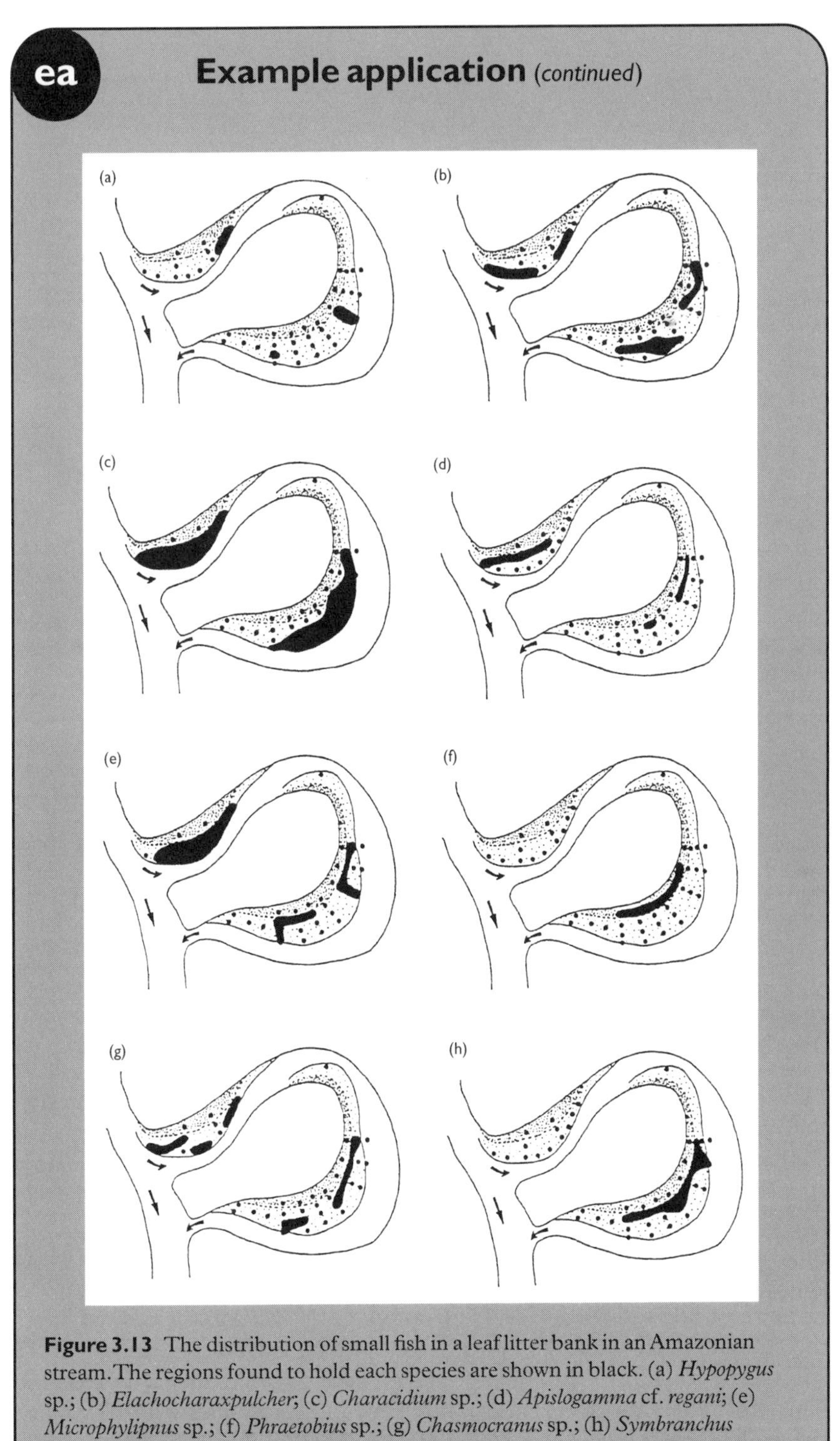

Figure 3.13 The distribution of small fish in a leaf litter bank in an Amazonian stream. The regions found to hold each species are shown in black. (a) *Hypopygus* sp.; (b) *Elachocharaxpulcher*; (c) *Characidium* sp.; (d) *Apislogamma* cf. *regani*; (e) *Microphylipnus* sp.; (f) *Phraetobius* sp.; (g) *Chasmocranus* sp.; (h) *Symbranchus marmoratus*. (After Henderson & Walker 1986.)

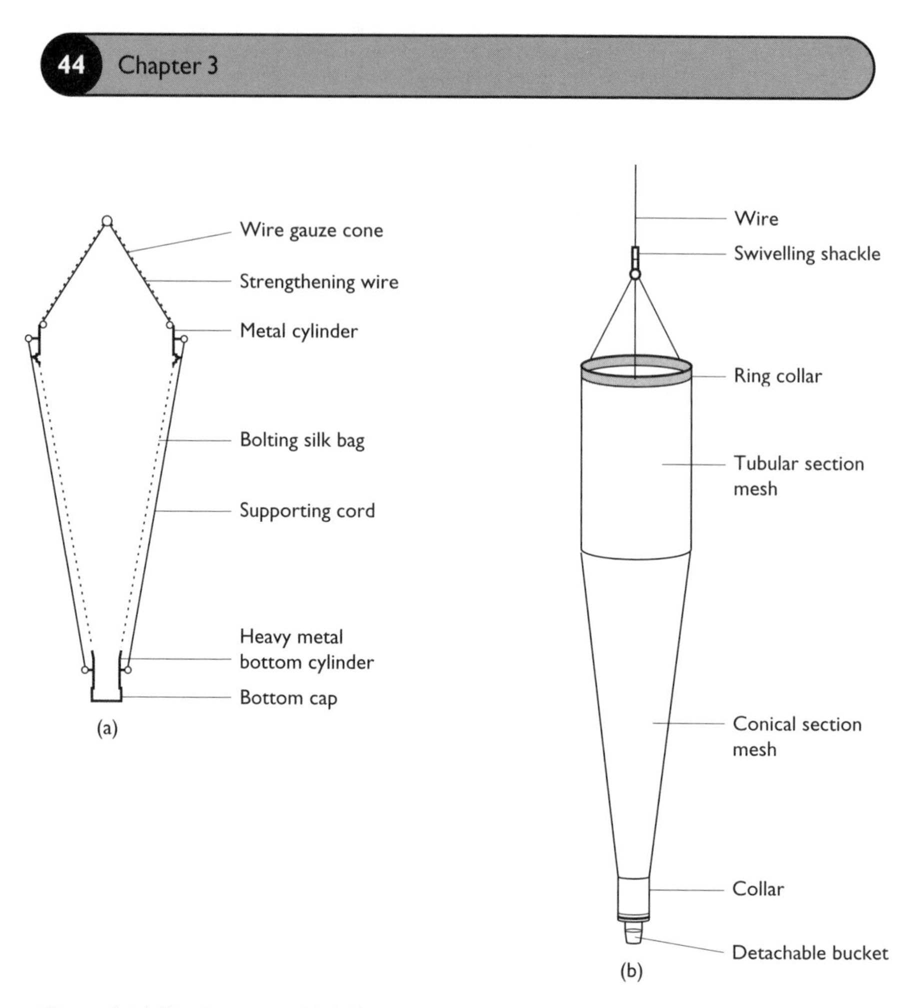

Figure 3.14 Plankton nets. (a) A birge cone tow-net suitable for freshwaters with beds of water plants. (b) A modified working party 2 net used for vertical tows from depths of less than 200 m up to the surface.

Sampling a known volume of water

Planktonic animals and plants are frequently sampled using plankton nets that are towed or pulled through the water (Figs 3.12, 3.14). In marine studies large plankton nets fitted with flow meters are pulled behind boats at speeds of between 3 and 5 knots. Such methods are unsuitable for small lakes and ponds, where the Patalas–Schindler volume sampler is a particularly useful quantitative sampler. The main body of the sampler is a perspex box with a capacity of 12 liters, with a large flap valve on the lower surface and a net attached to a sidewall (Fig. 3.15). The sampler is lowered to the desired depth then raised swiftly so that the valve closes, retaining the water sample. At the surface the box is turned on its side and its water drained out via the side net. This procedure can be repeated a number of times to filter a larger volume before the sample is collected from the net bucket.

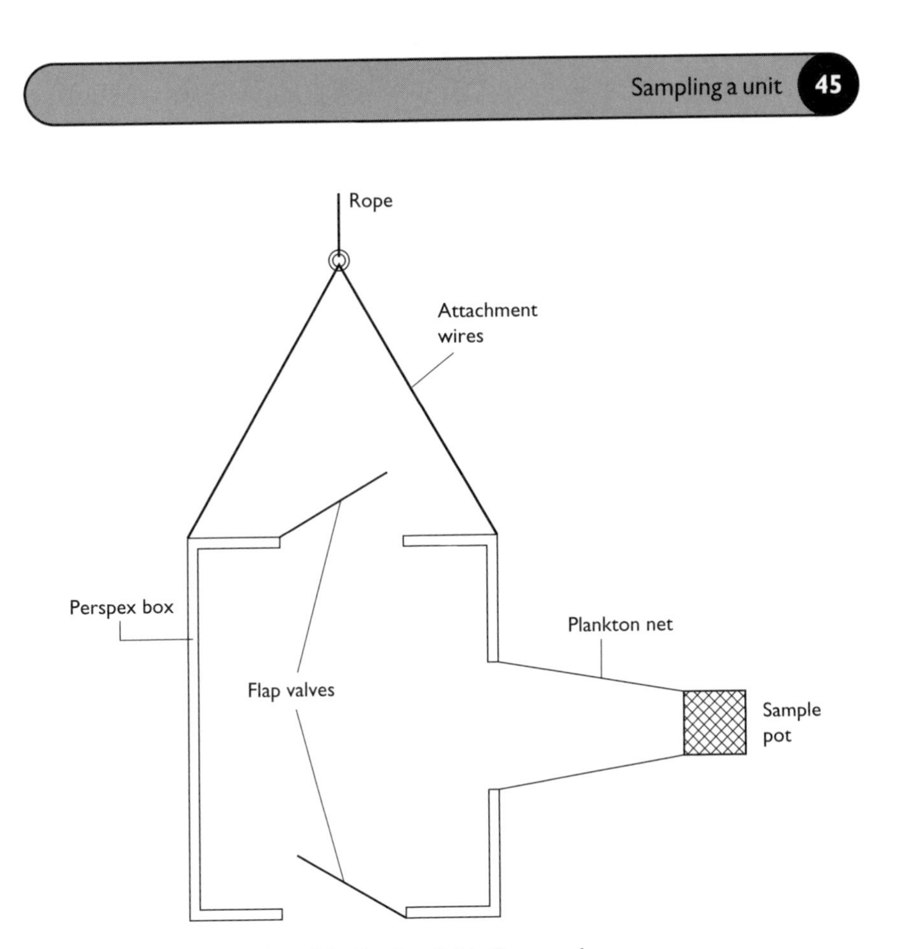

Figure 3.15 A cross-section of the Patalas–Schindler sampler.

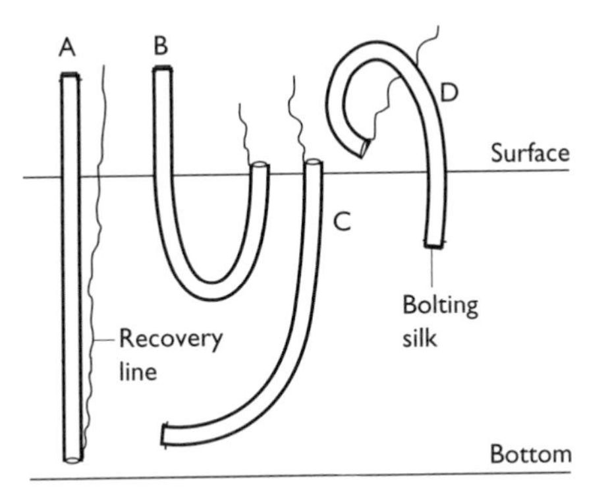

Figure 3.16 Perhaps the simplest sampling device possible – the littoral sampling tube showing the sequence of actions to collect a sample.

Perhaps the simplest practicable sampling device for zooplankton in vegetated littoral areas is the littoral sampling tube of Pennak (1962). This consists of a length of 60 mm diameter lightweight tubing. To one end is attached some plankton net mesh which can be removed, and to the open end

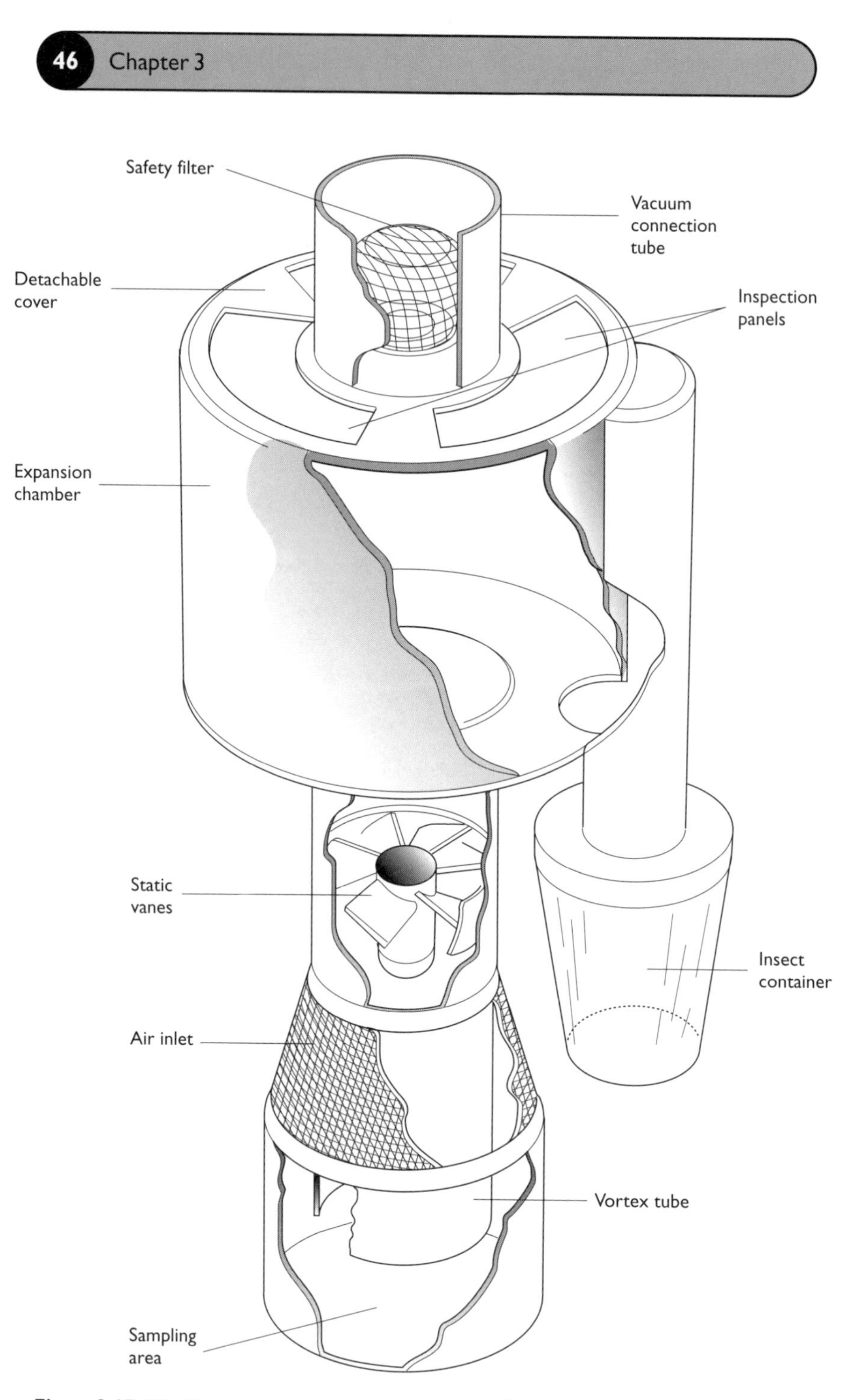

Figure 3.17 The Vortis suction sampler used for sampling arthropods in grassland and low herbage. (After Arnold 1994.)

a recovery line (Fig. 3.16). The sample is taken by lowering the open end of the tube vertically to the chosen depth ensuring that the mesh end remains in the air. Holding the mesh end steady the recovery line is used to raise the open end to the surface when the mesh end is released. The tube is then raised and the plankton retained on the mesh washed into a sample holder. Tubes up to 9 m long can be constructed and the method is applicable to habitats other than amongst aquatic vegetation, e.g. from under ice.

Using a suction sampler for grassland invertebrates

Based on the principles of domestic vacuum cleaners a number of types of suction sampler for animals living on or amongst the herbage down to the soil surface have been developed. Three widely used purpose-built commercial units are the D-Vac, Univac, and Vortis (Fig. 3.17), although the D-Vac is no longer produced. These machines work by enclosing a known area of grass from which the insects and other invertebrates are sucked. Cheap machines can be made by converting those designed for other material such as garden litter, and designs have been published that offer advantages of cost, weight and possibly sampling efficiency over the established commercial models (e.g. Stewart & Wright 1995). Suction sampling is only really effective in dry, upright vegetation less than about 15 cm high and the efficiency of suction samplers is affected by several factors, but of particular importance are nozzle wind speed and speed of enclosure of the sample area. When used under suitable conditions the results can be excellent.

You may need to use one of the extraction methods described elsewhere in this chapter to separate your catch from the plant debris in the sample.

Chapter Four

Mark–recapture methods for population size estimation

If you wish to estimate the size of the populations of animals that are easily captured without harm to the animal (or the biologist) and which live in populations of up to a few thousand individuals, mark–recapture techniques can be considered. The basic concept behind these methods is to take a sample from the population, count, mark and release them back into the population, then to take one or more further samples and calculate the size of the population from the number of marked and unmarked individuals present. In practice, a large number of experimental protocols and analytical schemes have been developed to cope with various complications such as death, tag loss, migration, or nonrandom mixing.

Populations are classified as either **open** or **closed**. The number of individuals in a closed population remains constant over the period of study whereas an open population changes because of some combination of birth, death, and emigration. If at all possible you should select a study site and a study time scale that will allow a closed population to be assumed. This is because you will avoid the need to estimate emigration in addition to population size; the greater the number of parameters that are to be estimated, the higher the proportion of the population that must be captured and marked. Three methods are introduced here, one applicable to closed and the other two to open populations. While simple methods might seem to lack sophistication, in practice, because they do not require that a large proportion of the population should be marked, these are often the methods of choice.

Mark–recapture methods cannot be used for the estimation of large populations because somewhere in the region of 20% of the population must eventually be captured and marked if the accuracy of the estimate is to be more than an order of magnitude approximation. Therefore, before embarking on a mark–recapture experiment you should consider: (i) the accuracy that is required; (ii) the possible size of the population; and (iii) the difficulty of making live captures. If you do not carefully consider these points you may make a costly mistake. It is often useful to undertake some simulations; these can be done by applying the method to beans in a bag and seeing how accurate the estimates are. A more sophisticated approach is to use simulation software such as SIMPLY TAGGING (www.irchouse.demon.co.uk). As a young ecologist I decided to estimate the population of brown shrimp in the Bristol Channel by mark–recapture. With a team of five helpers we were able to capture, mark, and release between 5000 and 10,000 shrimps per day, but subsequent analysis indicated that the population was in the order of 10^{10} individuals. To mark just 1% of this population would have required a team of 1000 working for 10 days! At the other extreme, animals with exceedingly low density populations may be impossible to study because insufficient numbers of individuals will be caught.

Important assumptions

There are four assumptions common to almost all mark–recapture techniques:

1 The marked animals are not affected either in behavior or life expectancy by being marked and the marks will not be lost.

2 The marked animals become completely mixed when released back into the population. Slow-moving or highly territorial animals are unsuitable for capture–recapture methods, as they do not move sufficiently to re-mix after marking.

3 The probability of capturing a marked animal is the same as that of capturing any member of the population; that is, the population is sampled randomly with respect to its mark status, age, and sex. Termed "equal catchability", this assumption is often difficult to achieve so you should consider what effect its violation will have on your estimate. It is clear that if some individuals cannot be caught the population will be underestimated. However, what you might be surprised to discover is that for some species, particularly small mammals, some individuals can be easier to catch than others and this will result in an underestimation.

4 Sampling must be at discrete time intervals and the actual time involved in taking the samples must be small in relation to the total time.

The Petersen–Lincoln estimator

If the population is closed, the probability of capture is constant, and the four common assumptions above are satisfied, then we can estimate the total population from the simple index used by Lincoln (1930):

$$\hat{N} = \frac{an}{r} \tag{4.1}$$

where $\hat{N}$ = the estimate of the number of individuals in the population, a = total number marked in the first sample, n = total number of individuals in the second sample, and r = total recaptures (individuals marked at time 1 and recaught at time 2).

An example calculation

On day 1 for 1 hour in the morning dragonflies around the perimeter of a pond were caught and marked using a marker pen. A total of 47 dragonflies were caught and marked on the first day. On the second day a total of 51 dragonflies were caught, of which 10 held marks from the first day. Therefore $a = 47$, $n = 51$, and $r = 10$. The estimated size of the population $= 47 \times 51/10 = 239.7 = 240$.

The above calculations are applicable to large samples where the value of r is fairly large; with small samples a less biased estimate based on a binomial approximation to the hypergeometric distribution is (Bailey 1951, 1952):

$$\hat{N} = \frac{n(a+1)}{r+1} \tag{4.2}$$

and an approximate estimate of the variance is:

$$\mathrm{var}(\hat{N}) = \frac{a^2(n_2+1)(n_2-r)}{(r+1)^2(r+2)} \tag{4.3}$$

A simple method for open populations – Bailey's triple-catch method

If the animals are marked on a series of two or more occasions, then an allowance may be made for the loss of marked individuals between the time of the initial release and the time when the population is estimated. The formulae for the calculation of the various parameters are relatively simple; Bailey (1951) gives the mathematical background and Bailey (1952) and MacLeod (1958) describe it. It has been used with a variety of insects and other invertebrates (e.g. snails (Woolhouse & Chandiwana 1990) and lobster (Evans & Lockwood 1994)).

For the purposes of illustration the sampling is said to take place on days 1, 2, and 3. In practice, the intervals between these sampling occasions may be of any length, provided they are long enough to allow good mixing of the marked individuals with the remainder of the population and not so long that a large proportion of the marked individuals have died. With large samples the population on the second day is estimated using:

$$\hat{N}_2 = \frac{a_2 n_2 r_{31}}{r_{21} r_{32}} \tag{4.4}$$

where a_2 = the number of newly marked animals released on the second day, n_2 = the total number of animals captured on the second day, and r = recaptures with the first subscript representing the day of capture and the second the day of marking; thus r_{21} = the number of animals captured on the second day that had been marked on the first, r_{31} = the number of animals captured on the third day that had been marked on the first. It is clear that we are really concerned with the number of marks, and hence the same animal could contribute to r_{31} and r_{32}. The large-sample variance of the estimate is:

$$\operatorname{var} \hat{N}_2 = \hat{N}_2^2 \left(\frac{1}{r_{21}} + \frac{1}{r_{32}} + \frac{1}{r_{31}} - \frac{1}{n_2} \right) \tag{4.5}$$

Where the numbers recaptured are fairly small there is some advantage in using "Bailey's correction factor" so that:

$$\hat{N}_2 = \frac{a_2 (n_2 + 1) r_{31}}{r_{21} r_{32}} \tag{4.6}$$

with approximate variance:

$$\operatorname{var} \hat{N}_2 = \hat{N}_2^2 - \frac{a_2^2 (n_2 + 1)(n_2 + 2)(r_{31} - 1) r_{31}}{(r_{21} + 1)(r_{21} + 2)(r_{32} + 1)(r_{32} + 2)} \tag{4.7}$$

The loss rate, which includes both the numbers dying and the numbers emigrating, is given by:

$$\gamma = \ln \left(\frac{a_2 r_{31}}{a_1 r_{32}} \right)^{\frac{1}{t_1}} \tag{4.8}$$

where t_1 = the time interval between the first and second sampling occasions. The dilution rate, which is the result of births and immigration, is given by:

$$\beta = \ln \left(\frac{r_2 n_3}{n_2 r_{31}} \right)^{\frac{1}{t_2}} \tag{4.9}$$

where t_2 = the time interval between the second and third sampling occasions. Both these are measures of a rate per unit of time (the unit being the units of t).

A more complex method for open populations – the Jolly–Seber model

The basic equation in Jolly's method is:

$$\hat{N}_i = \frac{\hat{M}_i n_i}{r_i} \tag{4.10}$$

where $\hat{N}_i$ = the estimate of population on day i, $\hat{M}_i$ = the estimate of the total number of marked animals in the population on day i (i.e. the counterpart of "a" in the simple Lincoln index), r_i = the total number of marked animals recaptured on day i, and n_i = the total number captured on day i.

The procedure may be demonstrated by the following example taken from Jolly (1965). (The notation is slightly modified to conform to the rest of this section.) It would be normal practice to undertake these calculations on a computer.

Table 4.1 The tabulation of recapture data according to the date on which the animal was last caught for analysis by Jolly's method. (After Jolly 1965.)

Day of capture (i)	Total captured (n_i)	Total released (a_i)	Day when last captured (j)												
1	54	54	*1*												
2	146	143	10	*2*											
3	169	164	3	34	*3*										
4	209	202	5	18	33	*4*									
5	220	214	2	8	13	30	*5*								
6	209	207	2	4	8	20	43	*6*							
7	250	243	1	6	5	10	34	56	*7*						
8	176	175	0	4	0	3	14	19	46	*8*					
9	172	169	0	2	4	2	11	12	28	51	*9*				
10	127	126	0	0	1	2	3	5	17	22	34	*10*			
11	123	120	1	2	3	1	0	4	8	12	16	30	*11*		
12	120	120	0	1	3	1	1	2	7	4	11	16	26	*12*	
13	142		0	1	0	2	3	3	2	10	9	12	18	35	*13*
	R_i =			80	70	71	109	101	108	99	70	58	44	35	

Table 4.2 Calculated table of the total number of marked animals recaptured on a given day bearing marks of the previous day or earlier. (After Jolly 1965.)

Day i	Day $i-1$												
	1												
2	**10**	*2*											
3	3	**37**	*3*										
4	5	23	**56**	*4*									
5	2	10	23	**53**	*5*								
6	2	6	14	34	**77**	*6*							
7	1	7	12	22	56	**112**	*7*						
8	0	4	4	7	21	40	**86**	*8*					
9	0	2	6	8	19	31	59	**110**	*9*				
10	0	0	1	3	6	11	28	50	**84**	*10*			
11	1	3	6	7	7	11	19	31	47	**77**	*11*		
12	0	1	4	5	6	8	15	19	30	46	**72**	*12*	
13	0	1	1	3	6	9	11	21	30	42	60	**95**	*13*
$Z(i-1)+1$ =	14	57	71	89	121	110	132	121	107	88	60		
	Z_2	Z_3	Z_4	Z_5	Z_6	Z_7	Z_8	Z_9	Z_{10}	Z_{11}	Z_{12}		

1 The field data are tabulated as in Table 4.1 according to the date of initial capture (or mark) and the date on which the animal was last captured. The columns are then summed to give the total number of animals released on the ith occasion ($= s_i$ of Jolly) subsequently recaptured (R_i), e.g. for day 7, $R = 108$.
2 Another table is drawn up (Table 4.2) giving the total number of animals recaptured on day i bearing marks of day j or earlier (Jolly's a_{ij}); this is done by adding each row in Table 4.1 from left to right and entering the accumu-

Table 4.3 The final table for a Jolly type mark and recapture analysis. (After Jolly 1965.)

Day i	Proportion of recaptures $\hat{\alpha}_i$	No. marked animals at risk $\hat{M}_i$	Total population $\hat{N}_i$	Survival rate $\hat{\phi}_i$	No. of new animals $\hat{B}_i$	Standard errors			Standard errors due to errors in the estimation of parameter itself	
						$\sqrt{\{v(\hat{N}_i)\}}$	$\sqrt{\{v(\hat{\phi}_i)\}}$	$\sqrt{\{v(\hat{B}_i)\}}$	$\sqrt{\{v(\hat{N}_i/N)\}}$	$\sqrt{\left\{V(\hat{\phi}_i)-\frac{\hat{\phi}_i^2(1-\hat{\phi}_i)}{\hat{M}_{i+1}}\right\}}$
1	—	0	—	0.649	—	—	0.114	—	—	0.093
2	0.0685	35.02	511.2	1.015	263.2	151.2	0.110	179.2	150.8	0.110
3	0.2189	170.54	779.1	0.867	291.8	129.3	0.107	137.7	128.9	0.105
4	0.2679	258.00	963.0	0.564	406.4	140.9	0.064	120.2	140.3	0.059
5	0.2409	227.73	945.3	0.836	96.9	125.5	0.075	111.4	124.3	0.073
6	0.3684	324.99	882.2	0.790	107.0	96.1	0.070	74.8	94.4	0.068
7	0.4480	359.50	802.5	0.651	135.7	74.8	0.056	55.6	72.4	0.052
8	0.4886	319.33	653.6	0.985	13.8	61.7	0.093	52.5	58.9	0.093
9	0.6395	402.13	628.8	0.686	49.0	61.9	0.080	34.2	59.1	0.077
10	0.6614	316.45	478.5	0.884	84.1	51.8	0.120	40.2	48.9	0.118
11	0.6260	317.00	506.4	0.771	74.5	65.8	0.128	41.1	63.7	0.126
12	0.6000	277.71	462.8	—	—	70.2	—	—	68.4	—
13	0.6690	—	—	—	—	—	—	—	—	—

lated totals. The number marked before time i that are not caught in the ith sample, but are caught subsequently (Z_i), is found by adding all but the top entry (printed in bold) in each column. Thus Z_7 is given by the figures enclosed in Table 4.2. The figures above the line, i.e. the top entry in each column, represent the number of recaptures (r_i) for the day on its right, e.g. $r_7 = 112$.

3 Then the estimate of the total number of marked animals at risk in the population on the sampling day may be made (Table 4.3):

$$\hat{M}_i = \frac{a_i Z_i}{R_i} + r_i \tag{4.11}$$

Thus $M_7 = 243 \times 110/108 + 112 = 359.50$.

4 The proportion of marked animals in the population at the moment of capture on day i is calculated and entered in the final Table 4.3 using:

$$\alpha_i = \frac{r_i}{n_i} \tag{4.12}$$

For example:

$$\alpha_7 = \frac{112}{250} = 0.448 \tag{4.13}$$

5 The total population is then estimated for each day using:

$$\hat{N} = \frac{\hat{M}_i}{\alpha_i} \tag{4.14}$$

6 The probability that an animal alive at the moment of release of the ith sample will survive until the time of capture of the $i + 1$th sample is calculated using:

$$\hat{\Phi}_i = \frac{M_{i+1}}{\hat{M}_i - r_i - a_i} \tag{4.15}$$

Survival rates should be less than or equal to 1; estimates slightly over 1 can happen, but "rates" much above 1 indicates an error. Frequently it will be found that the marks of one occasion have been lost or were not recognized.

7 The number of new animals joining the population in the interval between the ith and $i + 1$th samples and alive at time $i + 1$ is given by:

$$\hat{B}_i = \hat{N}_{i+1} - \hat{\Phi}_i(\hat{N}_i - n_i + a_i) \tag{4.16}$$

This may be converted to the dilution rate β:

$$\frac{1}{\beta} = 1 - \frac{\hat{B}_i}{\hat{N}_{i+1}} \tag{4.17}$$

Methods for marking animals

Individual and batch marking

Marking techniques are often classified as either group (batch) or individual (tagging) techniques. An example of an individual tagging method would be the attachment of a numbered ear tag to a mammal. Methods such as fin clipping, branding or staining are often used to mark all individuals caught at one time in a similar manner and are therefore batch marking methods. Batch marking methods are used with either the Lincoln index or Bailey triple-catch methods. The Jolly–Seber method requires sampling at a number of points in time and the creation of a record of the occasions when an individual was caught and therefore requires individual tagging.

Necessary criteria for a useful marking method

The chosen method of marking must seek to meet the following list of requirements:

1 Marking should not affect growth, longevity, or behavior of the animals. In practice, this is often impossible to prove and is generally unlikely to be the case because of the distress and damage caused during capture and marking.

2 Marking should not affect vulnerability to predation. An animal's natural camouflage may be lost making it more or less liable to predation. This effect is difficult to assess; it can to some extent be avoided by marking in inconspicuous places.

3 The mark must be durable. Remember that invertebrates will lose marks on their cuticle when they moult.

4 The mark must be easily detected.

5 The amount of effort invested into a marking program and the choice of marking method need to be related to the percentage of recoveries that can be expected. A high cost per individual will be justified where the recovery rate is high or the animals have a special significance.

6 Generally, marks must be easily applied in the field and without the need for anesthetic. Ethical aspects of the distress and harm caused to the animals need to be considered (Putman 1995).

7 Consideration may also have to be given to the effect of preservatives on the mark and its toxicity, appearance and safety if the marked animals can be seen, caught or eaten by the public.

Handling techniques

Handling can be stressful, even fatal, and should be carefully considered. The marked animals will not behave normally if handling is poor. If the animals are to be marked by spraying or dusting this can be done while they are still active, either in the field or in small cages. But, for marking with paints, especially precise spotting, and for most methods of labeling and mutilation, it is necessary for the animal to be still. Insects, crustaceans, molluscs, birds, and small mammals can often be held in the hand or between the fingers; where this is impossible the animal must be held by some other method, anesthetized or chilled. Fish can be placed between two boards fixed to give a V-shaped cross-section. Invertebrates may be held still under a net or with a hair or may be held by suction.

If none of these methods can be used or if the insects are so active that they cannot even be handled and counted they may have to be immobilized. Chilling at a temperature between 1 and 5°C is probably the best method. Alternatively, as a last resort, an anesthetic may be employed: carbon dioxide is often used and is easily produced from "dry ice" or bought in cartridges designed to carbonate drinks or obtained from fire extinguishers. Working on the earwig, *Forficula*, Lamb and Wellington (1974) found that although activity may be quickly resumed after carbon dioxide anesthesia of less than 1 minute, normal behavior could be affected for many hours; they recommend a 24 hour recovery period. Other anesthetics are ether, chloroform, nitrogen, and nitrous oxide. Fish are one group where the use of anesthetics is advised (Laird & Stott 1978). MS 222 and benzocaine are the two most commonly used.

Release

Only apparently healthy, unharmed individuals should be released after marking. Animals frequently show a high level of activity immediately after release and this may lead to excess dispersal or losses from predators. Two approaches can be used to minimize this effect. If the animal has a marked periodicity of activity (i.e. it is diurnal or nocturnal) then it should be released during the inactive period. Insects that are active at most times of the day may be restrained from flying immediately after release by covering them with small cages.

For nonterritorial animals the release sites should be chosen carefully and should be scattered throughout the habitat, as it is essential that the marked animals mix freely with the remainder of the population. The extent of the re-mixing may be checked, to some degree, by a comparison of the ratio of marked to unmarked individuals in samples from various parts of the habitat. Frequently, marked crustaceans and fish have adapted their coloration to the conditions within the holding tanks so that they have no natural camouflage when released.

ea Example applications

Intertidal and brackish pools

A number of crustaceans and molluscs living in intertidal pools can be estimated by mark–recapture techniques. The assumption of random mixing is difficult to achieve using molluscs but is probably appropriate for prawns and shrimps that are to be found in almost all marine littoral habitats and saltmarsh drainage ditches. Shrimps can be easily captured using baited traps; however, because it is quicker, it is usually best to sample using a hand net allowing the experiment to be undertaken over a single low tide or within one day. This in turn will allow the Petersen–Lincoln closed population estimate to be used. Shrimps can be marked by placing a dot of cyanoacrylate adhesive (Superglue) on the carapace. The glue can be made more visible by mixing it with a pigment such as talcum powder. The resulting white dot on the carapace will harden in or out of water. The adhesive will harden almost immediately if only a small quantity is exposed to the air, but a quantity sufficient to half-fill a solid watch glass mixed with powder will stay liquid for long enough to allow a large batch of shrimps to be marked. A suitable protocol is to catch as many shrimps as possible for a fixed time of say 20 minutes and to mark, count and release them back into the pool. Wait for about 1 hour by which time the population should be well mixed, and sample for a second time and count the number of marked and unmarked shrimps in the second sample.

Fish in ponds

Ponds offer an ideal habitat for the use of the Petersen–Lincoln closed population method to estimate the populations of small fish. For fish such as minnows and sticklebacks the best sampling method is usually to use traps (see page 80). It is not usually necessary to bait the traps, although it is useful for some species. Alternatively, some species can be caught using a small seine net, possibly only 4 or 5 yards (1.2–1.5 m) long with a fine mesh of only 5 mm, that can be handled by just two people. A suitable batch marking method is fin clipping normally applied to the tail fin. If a small notch is cut or the tip is cut off the tail with a pair of scissors no long-term harm will be done to the fish. Take great care to avoid hurting their skin, and always handle fish with wet hands. It may be advisable to use anesthetics (see "Methods for marking animals" above). A suitable sampling protocol is as follows:

Day 1. In the morning place traps around the perimeter of the pond. In the afternoon lift the traps, mark the fish, record the number marked and quickly return them to the pond close to the point of capture. It is often useful to place the fish in a recovery tank prior to release to ensure they are fully recovered; struggling, exhausted fish are highly vulnerable to predators.

Day 2. In the morning reset the traps. They do not need to be in the same localities as used on day 1. In the afternoon, lift the traps and count the number of marked and unmarked individuals and return them quickly to the water. The equations given earlier in this chapter can then be used to calculate population size.

Estimating the populations of aquatic beetles and bugs in ponds

Most aquatic Coleoptera and Hemiptera can fly and therefore one cannot be sure that the population of a pond will be closed to emigration and immigration if a mark–

(continued on page 58)

ea Example applications *(continued)*

recapture experiment lasts more than 1 day. This makes these groups appropriate for the application of the Bailey triple-catch method. These insects can be easily sampled with a hand net and almost all ponds hold a number of species. It is best to study the populations of the larger beetles or bugs, as these normally do not have very large populations per pond and are more easily handled without being harmed. Once a species or higher taxonomic group has been chosen for study, a marking method must be decided upon. For aquatic insects the most reliable marking method is probably a mutilation technique such as scratching a mark on the elytra with a dissecting needle. The Bailey triple-catch requirement that different marks be applied to animals released on day 1 and day 2 can be achieved by marking the left and right elytra respectively. While it would be appropriate to sample a number of times in 1 day and use a closed population method, these groups are probably useful to demonstrate an open population method for which it is best to sample at intervals of 1 or more days. This will give time for appreciable population change to occur – some aquatic insects only fly at night. A possible schedule is as follows:

Day 1. In the morning use a pond net to catch, mark, count, and release the insects. Take care to ensure all the animals carry the same mark.

Day 2. In the morning resample the pond. Examine all the individuals captured for day 1 marks, record the number found, add a unique mark for day 2, record the total number captured and release the insects.

Day 3. In the morning resample the pond and examine all the animals captured and note the number baring marks from day 1 only, day 2 only, days 1 and 2, and no marks. You now have the data needed to calculate population size and loss rates using the Bailey triple-catch method.

Butterflies

Given a discrete area of habitat such as a clearing in a forest, a field or area holding the host plant, or perhaps even a group of flowering shrubs such as buddleia which are particularly attractive to butterflies, a mark–recapture experiment can be undertaken. While some tropical butterflies are best trapped, many species can be caught using a butterfly net. While wing clipping can mark butterflies, such mutilation may not be advisable and they are probably best marked with a paint or stain. Suitable stains are Eosine, Orange G or Congo red dissolved in alcohol, which can be painted onto a wing. A closed or an open population estimation protocol may be appropriate.

Ground-living beetles and terrestrial isopods

Carabid beetles are particularly good subjects for mark–recapture experiments as the carapace is easily marked with a spot of acrylic paint. The most suitable sampling method is pitfall trapping with the traps set overnight. You must take care to ensure that the traps will not fill with water if it rains by giving them roofs and drainage holes. The problem is to delimit an area of study. The most desirable situation is to study the population of an "island" population such as a clearing in a forest or a woodland coppice in open farmland. The pitfall traps will need to be dispersed over the study area. As it is unlikely that a population would be closed, it is probably best to use the Bailey triple-catch method over 3 or more days with the exception that sampling occurs

(continued)

ea

Example applications *(continued)*

overnight. If time allows, these animals are suitable candidates for a Jolly–Seber method.

Small mammals

Some of the most popular subjects for mark–recapture experiments are small mammals. These animals can be captured alive using a Longworth or similar trap. The traps must be supplied with nesting material and food or the captured animals may suffer or even die. Care must be taken to ensure that you know how to handle the study animals with confidence as mammals are easily injured or stressed; further, many can bite and some may transmit disease to the handler. One of the best methods of marking, which is capable of producing individual marks for life, is ear tattooing with tattooing tongs. Other possible methods include clipping small bits from the ear under local anesthetic, freeze branding, and tagging.

Small mammals are territorial and generally live in social groups with a distinct hierarchy. These features suggest that great care must be taken with experimental design if a reasonable population estimate is to be obtained. It is likely that the two sexes will need to be estimated separately as they are frequently observed to enter traps at different rates. This will require the sex of each captured animal to be noted and the males and females used to create independent data sets for analysis. For most species the traps are set overnight and it may also be necessary to deploy them without the trap-door mechanism activated for a number of nights to allow the animals to overcome their suspicions and enter the traps. To estimate the population it would be usual to set at least 20–50 traps each night. The traps need to be spread over the study area. However, it is probably not best to arrange the traps into a regular grid; rather they should be placed in locations that experience has shown are attractive to the mammals.

To obtain sufficient data, sampling and marking will normally need to be undertaken for a number of days and this in turn points to the use of the Jolly–Seber method of analysis.

Chapter Five

Distance sampling methods for population estimation

Counting the number of sightings forms the basis for estimating density for many animal and plant groups. This is particularly the case for large or easily seen organisms such as birds, large grassland mammals, whales, and showy, active insects such as butterflies. While it may be possible to count organisms from a suitable vantage point or while moving along a transect, the count can only be converted to a density estimate if the area scanned can be estimated. This simple approach is often difficult to undertake for two reasons: firstly it may not be possible to estimate sufficiently accurately the area scanned and, secondly, not all of the organisms present may have been spotted. Distancing sampling methods have been developed to allow for these problems by assuming that the likelihood that an individual will be observed will decline in a mathematically definable way with distance. Another approach when estimating the populations of easily caught animals to is use removal sampling methods. For these, the decline in the rate of capture as the population is removed is used to estimate the original number present.

A second category of distance sampling methodologies, called plotless estimators, is used for plants and sessile or slow-moving organisms such as corals or sponges. This is based on the measurement of distances between individuals or from fixed points to individuals.

Census methods

If it proves possible to count all of the individuals, n, within a known area, a, then this is termed a census and the estimated density, $\hat{D}$, is simply:

$$\hat{D} = \frac{n}{a} \quad (5.1)$$

Counting often requires the observer to move over the census area and thus favors the use of strip transects (long, thin quadrats) of length, L, and width, $2w$, along which the observer moves in a straight line. Strips have been surveyed by foot, car, boat, or aircraft. The strip width must be determined at the outset; if this cannot be done, or animals may be missed, then line transect methods (see below) should be used.

Some animals concentrate at particular seasons or times of the day when a census becomes possible. Allsteadt and Vaughan (1992) were able to census caiman during the dry season by counting the reflective eyes by torchlight at night; each animal appears as two bright red dots and the size (but not the intelligence) can be estimated by the distance between the eyes. Technological developments that improve detection rates are also making census methods more feasible. Naugle *et al.* (1996) were able to detect white-tailed deer using an aerial infrared imaging system. Image-intensifying night sights are now frequently used to observe mammals at night.

Point and line survey methods

Indices of abundance using transects

These methods are frequently used in conservation studies where minimizing cost is important and when the objective is to detect any change in abundance rather than absolute magnitude. A typical example applied to insects is the study by Thomas (1983) on a number of species of British butterflies. The protocol was:

1 Undertake a preliminary inspection to define the extent of the butterfly colony: this is termed the flight area, A.

2 Mark out a zig-zag trail of length L m which should either exceed 1000 m in length or be long enough to record 40 individuals.

3 Walk the trail and count individuals, N, seen within an imaginary box from 0 to 5 m ahead with a fixed width which varies with species, but is typically 4–6 m.

4 Calculate the index as:

$$p = \frac{NA}{L} \tag{5.2}$$

Thomas (1983) showed that this abundance index was correlated to the absolute population density as determined by mark–recapture (see Chapter 4) and could be converted to an estimate of absolute abundance by regression analysis.

Methods based on flushing

The simplest approach is to estimate the number disturbed from a known width of habitat. Under certain circumstances, e.g. seabird colonies, all individuals may be flushed and the problem is simply obtaining an accurate count (Bibby *et al.* 1992). In most habitats only a proportion of the animals are disturbed. If this is constant then the numbers themselves give an index of absolute population and, incorporating the proportion and the area covered, an estimate of absolute population density. If the proportion flushed is unknown or varies, only a relative measure of availability is given.

Hayne (1949) developed the first estimator of density for a flushing experiment that could claim to be robust for studies on bird populations. This simple method was designed to estimate the numbers of grouse, which it was assumed would flush as the observer came within a certain radius, *r*, of the bird. The observer moves through the habitat along a transect noting the radial distance at which each bird is flushed. The density estimate, $\hat{D}_H$, is then given by:

$$\hat{D}_H = \frac{n}{2L}\left(\frac{1}{n}\sum\frac{1}{r_i}\right) \tag{5.3}$$

where n is the total number of animals counted, L the transect length, and r_i the sighting distance to the ith animal.

The approximate variance of this estimate is:

$$\text{var}(\hat{D}_H) \approx D_H^2\left[\frac{\text{var}(n)}{n^2} + \frac{\sum\left(\frac{1}{r_i} - R\right)^2}{R^2 n(n-1)}\right] \tag{5.4}$$

where R is the mean of the reciprocals of the sighting distances.

More than six other estimators have been developed (Gates *et al.* 1968; Gates 1969). An unbiased estimator with minimal variance is considered to be that of Gates (1969) and Kovner and Patil (1974), although some others may be more robust. This estimator is:

$$\hat{D} = [2n - 12L\bar{r}] \tag{5.5}$$

where n = number of animals sighted, L = length of the line transect, and $\bar{r}$ = average radial distance over which the observer encountered animals. The assumptions are made that encounter distances (r) follow a negative exponential distribution and that the flushing of one animal will not affect another.

The estimation of each r can, of course, be a major source of error. The observer should seek to fix the position where the animal was seen and then measure the distance; a range finder may be helpful (Bibby *et al.* 1992).

The transect route may simply follow an existing path and this may be satisfactory for relative estimates. A more acceptable basis for absolute population estimates will be provided if the single route is replaced with a number of shorter, straight transects each with starting point and direction determined randomly. If the habitat has different zones each should be traversed separately (Bibby *et al.* 1992).

Line transect methods – the Fourier series estimator

Line transect methods have been developed for situations when it is not possible to count all the animals within a strip transect. The methods are based on the idea that only animals lying on the center line of the strip transect along which the observer moves will be certain to be detected and that the probability of detection will fall with perpendicular distance from this line. This was also the case with the Haynes and Gates estimators discussed above, but they made strong assumptions about the shape of the detection function (rectangular and negative exponential respectively) that will often not hold. The techniques here have been reviewed in detail by Buckland *et al.* (1993).

For these methods it is assumed that:

1 Objects on the line are always detected.

2 The observer does not influence the recorded positions. For mobile animals, the position must be that prior to any response to the presence of the observer: the theory has been developed under the assumption that the objects are immobile, but slow movement in relation to the observer creates little inaccuracy.

3 Distances and/or angles are measured accurately.

The basic field procedure is for the transect route to be a straight line of length L, randomly placed with respect to the animals or plants to be counted, and the perpendicular distance to each detected object of interest, x, recorded. In practice a number of lines, arranged as a regular grid, and randomly placed in the study may be used.

Further, it is usually easier to record the sighting distance, r, which is the distance from the observer to the object, and the sighting angle, θ, which is the angle of the object from the transect line and calculate x rather than measuring x directly (Fig. 5.1).

If all n objects in a strip of length L and width $2w$ are counted, then the estimated density is:

$$\hat{D} = \frac{n}{2wL} \tag{5.6}$$

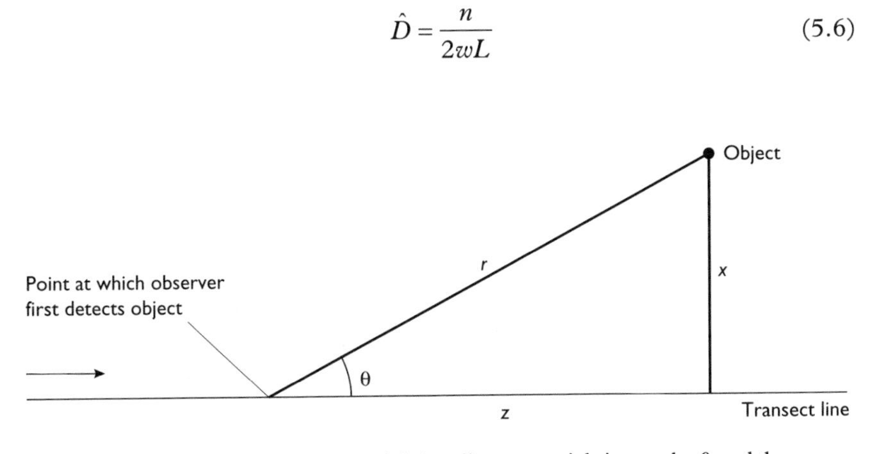

Figure 5.1 The relationship between sighting distance, r, sighting angle, θ, and the perpendicular distance, x, from the transect line.

If a proportion, P, of the animals present are detected then the equation becomes:

$$\hat{D} = \frac{n}{2wLP} \quad (5.7)$$

Line transect methods use the distribution of the perpendicular detection distances to estimate P. Figure 5.2 shows a typical histogram of the decline in the number of observations with perpendicular distance. A detection function $g(x)$, defined as the probability of detecting an object at distance x, is fitted to these data. Given assumption **1** above, $g(0)$, the probability of detecting an object lying on the line, is assumed to equal 1. The probability of detecting an object within a strip of area $2wL$, P, is:

$$P = \frac{\int_0^w g(x)\mathrm{d}x}{w} \quad (5.8)$$

which on substitution gives:

$$\hat{D} = \frac{n}{2L\int_0^w \hat{g}(x)\mathrm{d}x} \quad (5.9)$$

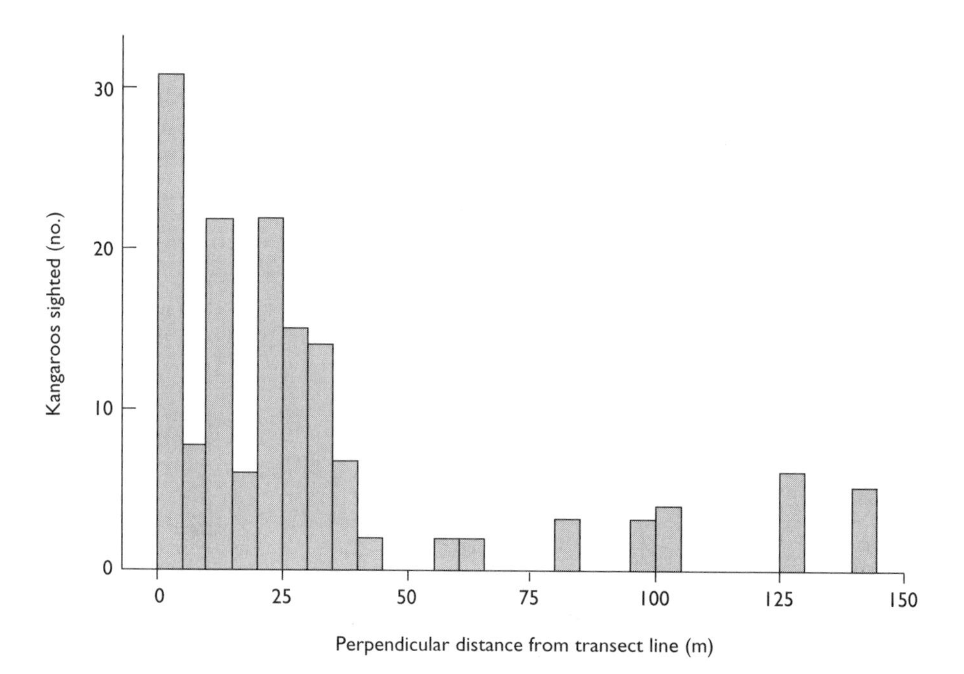

Figure 5.2 A typical histogram of the decline in the number of observations with perpendicular distance from the transect line. The data is for kangaroos. (After Coulson & Raines 1985.)

As it is assumed $g(0) = 1$, the probability density function (pdf) evaluated at $x = 0$ is:

$$f(0) = \frac{1}{\int_0^w g(x)\mathrm{d}x} \tag{5.10}$$

and thus the general estimator of density is often expressed as:

$$\hat{D} = \frac{n\hat{f}(0)}{2L} \tag{5.11}$$

The key issue is to estimate the detection function and this is done using a computer. Many different mathematical functions could be used to describe the decline in the detection of plants or animals with distance, but it has been found that a Fourier series is particularly appropriate and is fitted using programs such as DISTANCE (www.ruwpa.st-and.ac.uk/distancebook/) or the less sophisticated DENSITY FROM DISTANCE (www.irchouse.demon.co.uk).

Point transects

Instead of traversing a transect the observer may move to a number of fixed points and record the distance, r, to individual animals. These methods are almost only used for bird surveys where the patchy suitability of the habitat to the birds may make transects inappropriate because they cut across a number of habitat types. Point transects are often easier to undertake because the observer needs only estimate distance and markers may be placed in advance to aid the estimation of distance. Population density is given by:

$$\hat{D} = \frac{n\hat{h}(0)}{2\pi k} \tag{5.12}$$

where n is the number of animals observed, k the number of point transects undertaken, and $\hat{h}(0)$ the slope of the probability density function of detection distances evaluated at zero distance. In similar fashion to what was described for the line transect method, the central problem is to estimate $\hat{h}(0)$. This is normally accomplished using specialist computer programs such as DISTANCE or DENSITY FROM DISTANCE.

In the case of a half normal detection function the maximum likelihood estimator for density has the particularly simple form:

$$\hat{D} = \frac{n^2}{2\pi k \sum_{i=1}^{n} r_i^2} \tag{5.13}$$

An even simpler approach is to count the number of organisms within and beyond a set distance from the observation point. This approach is often used for the estimation of bird density (see example below). The observer counts the number of birds, N, within a set radius, r, and those birds observed beyond the distance r, N_2. The density is then estimated as:

$$D = \frac{\ln\left(\frac{N}{N_2}\right)N}{\pi r^2} \tag{5.14}$$

Normally a number of point samples is taken and the counts from all the points are pooled. When this is done you must remember to divide by the total number of sample points (see the example below "Point transects for birds"). The book *Bird Census Techniques* (Bibby *et al.* 1992) gives a good account of the application of line and point transects for the estimation of bird density.

Plotless density estimators

In their simplest form these methods are based on the premise that if the individuals of a population are randomly distributed then the population density can be estimated from either the distance between individuals (nearest-neighbor methods) or from randomly chosen points to the nearest individual (closest-individual methods).

Although widely used with vegetation, especially trees, these approaches have been little used by zoologists. They are however particularly appropriate for animals living or nesting in fairly uniform, two-dimensional habitats, such as muddy or sandy littoral zones.

Whilst the nonrandomness of many populations may preclude using the basic equations for exact population estimation, they do have the potential to provide an order of magnitude estimate.

A complete review of plotless density estimators is provided by Engeman *et al.* (1994). They used simulated data to assess the utility and robustness of the various estimators with different spatial patterns and different sample sizes. Their major conclusions are incorporated in the accounts below.

Closest individual or distance method

A point is selected at random and then by searching round it in concentric rings the closest individual is found and its distance (X) from the point measured. The process should be continued to find the second, third, . . . ith closest individual. Referred to by Engeman *et al.* (1994) as an "ordered distance estimator", they recommend using the third as being the most practical, as did Keuls *et al.* (1963) who used it for estimating the density of the snail *Limnaea truncatula*. The estimate of density, $\hat{D}$, is:

$$\hat{D} = \frac{sr - 1}{\pi \sum_{i=1}^{s} X_i} \tag{5.15}$$

where r = the rank of the individual in distance from the randomly selected point, e.g. for the nearest neighbor $r = 1$, for the second nearest 2 . . . ; X = the distance between the randomly selected point and the individual, and s = the number of random sampling points.

The variance of $\hat{D}$ is given by Seber (1982) as:

$$\text{var}[\hat{D}] = \frac{\hat{D}^2}{(sr - 2)} \quad (5.16)$$

A somewhat different approach, the variable area transect was developed by Parker (1979) and was considered by Engeman *et al.* (1994) to be the simplest and most practical. The search area is a fixed width strip of width w. Commencing from a random point the area is searched until the ith individual is found, when i is the number of individuals searched for from each random point and l is the length searched from the random point to the ith individual, then:

$$D = \frac{(si - 1)}{w \sum l} \quad (5.17)$$

Nearest-neighbor methods

In these methods a plant or animal is selected at random and searching is continued until another, the nearest neighbor, is encountered. The distance between the two (r) is measured. As with the closest individual method there are variants that use the second to ith nearest neighbor. The basic expression is due to Clark and Evans (1954):

$$\hat{D} = \frac{1}{4\bar{r}^2} \quad (5.18)$$

where $\hat{D}$ = density per unit area and $\bar{r}$ = mean distance between nearest neighbors. In this simple form the method will give poor results if the distribution is nonrandom (Engeman *et al.* 1994) and a number of superior estimators has been devised. These are based on the separate consideration of each quadrat in the circle surrounding the original animal (angle order estimators) or on a combination of information on the closest individual and nearest neighbor (methods of Kendal and Moran); however, the algorithms for their calculation are extremely complex (Engeman *et al.* 1994).

ea Example applications

Plotless methods

These population estimators can be used for a wide variety of plants and fungi, but take care to consider if the distribution can be considered random or if as is often the case highly clumped. Other suitable subjects are ant or termite nests (these may be regularly distributed) and worm casts on open ground or beaches.

Line transect techniques

A wide variety of plants, animals and even inanimate objects can be estimated by line transect methods. These techniques are particularly appropriate for large mammals, birds and active insects such as butterflies and dragonflies. The flushing methods can be used for some birds and insects.

Point transects for birds

These techniques are most appropriate for bird surveys where it is best for the observer to stay in one position so that the animals to be counted will not be alarmed.

Blue tit abundance was estimated by point sampling. A 2 km^2 area of scrub and open woodland was divided into 10 approximately equal areas and within each area a locality was chosen from which to undertake point counts. These localities were not chosen at random, but were the most suitable places from which to count the birds. On the day of the survey each point count was conducted for 5 minutes and the number of tits observed within and beyond 30 m of the survey point were counted. In total 45 tits were observed with 27 within the 30 m band. The density is calculated using the two-band method described at the end of "Point and line survey methods" above:

The total number of tits counted, $N = 47$.

The number of tits observed beyond 30 m radius, $N_2 = 45 - 27 = 18$.

The radius, $r = 30$ m.

The number of replicates = 10.

Using the equation in "Point and line survey methods" above but dividing by 10 for the number of point samples undertaken:

$$\text{Density} = \ln(47/18) \times 47/(10 \times 3.14 \times 30 \times 30)$$
$$= \ln(2.6) \times 47/28{,}260$$
$$= 0.00159 \text{ individuals per m}^2.$$

Estimating population size by removal sampling

The principle of removal trapping or collecting is that the number of animals in a closed*, finite, population will decline as will the catch per unit effort if captured animals are removed each time the population is sampled. These methods are particularly appropriate to small, isolated populations for which the probability of capturing each individual on each sampling occasion is

*A closed population is not open to immigration or emigration.

greater than 0.2. Removal methods are particularly appropriate for the estimation of fish populations in streams and ponds that are sampled by electric fishing. This approach does not require that the animals be killed and, if possible, you should arrange for them to be held captive and released at the end of the estimation.

The basic procedure is to undertake *s* distinct periods of trapping and to record the number caught for the first time on each occasion, u_s. The various approaches for calculating *N*, the total population size, are based on either maximum likelihood (ML) or regression methods. Given constant sampling effort (e.g. the same number of traps on each occasion or the same manpower used on each sweep of the river), ML methods are superior and will be described in detail below. ML estimates of *N* with a constant probability of capture model were first published by Moran (1951) and developed by Zippin (1956). A more general model allowing variable probabilities of capture was presented by Otis *et al.* (1978). Regression methods are still used in studies where the sampling effort is variable because it is not under the control of the researcher. For example, in fisheries research the number of fish landed varies with the activity of the fishing fleet. Methods for fisheries research are given in Ricker (1975) and Seber (1982) and will not be considered further here. When designing a removal trapping experiment every effort should be made to maintain a constant sampling effort on each trapping occasion. Further, it is important to plan to achieve a constant probability of capture. If different sexes or ages of individual differ in their probability of capture then estimate each of these components of the population separately.

Maximum likelihood method with constant probability of capture

To use this method the following conditions must be satisfied:

1 The catching or trapping procedure must not lower (or increase) the probability of an animal being caught. For example, the method will not be applicable if insects are being caught by the sweep net and after the first collection the insects drop from the tops of the vegetation and remain around the bases of the plants, or if the animals are being searched for and the most conspicuous ones are removed first.

2 The population must remain stable during the trapping or catching period; there must not be any appreciable natality, mortality, or migration. The experimental procedure must not disturb the animals so that they flee from the area.

3 The population must not be so large that the catching of one member interferes with the catching of another. This may be significant in vertebrate populations where a trap can only hold one animal.

4 The chance of being caught must be equal for all animals. This is the most serious limitation in practice. For example, some individuals of an insect population may never visit the tops of the vegetation and so will not be exposed to collection by a sweep net. Some small mammals may be "trap-shy". When electric fishing, smaller individuals are more difficult to stun.

If the probabilities of capture fall off with time, the population will be underestimated, but, if the animals become progressively more susceptible to capture, the population will be overestimated. Changes in susceptibility to capture will arise not only from the effect of the experiment on the animal, but also from changes in behavior associated with weather conditions or a diel periodicity cycle.

The theory

The expected number captured on each sampling occasion $E(u_s)$ is:

$$E(u_s) = N(1-p)^{s-1}p \tag{5.19}$$

where p is the probability of capture on each sampling occasion. Thus for the first sampling occasion the expected number caught is Np and for the second $N(1-p)p$ and so on.

Numerical iterative techniques are required to find the ML value for N and are best undertaken on a computer using programs such as POPULATION ESTIMATION BY REMOVAL SAMPLING (Pisces Conservation Ltd, www.irchouse.demon.co.uk\softremoval.html).

The ML estimates for N and p are given by solving the equations:

$$\hat{N} = \frac{T}{(1-\hat{q}^s)} \tag{5.20}$$

and

$$\frac{\hat{q}}{\hat{p}} - \frac{k\hat{q}^s}{(1-\hat{q}^s)} = \frac{\sum_{i=1}^{s}(i-1)u_i}{T} = R \tag{5.21}$$

where T is the total caught over all k samples and $q = 1 - p$. First R is calculated and then q is estimated numerically. This value is then used to estimate N. Using the method of Zippin (1956, 1958), it is shown below how the calculations can be undertaken for $k = 3$ to 7 without the aid of a computer. Zippin's procedure is well illustrated using the example taken from Southwood and Henderson (2000) and presented in Table 5.1.

Table 5.1 Example data from a removal trapping experiment.

	Sampling occasion					
	1st (u_1)	2nd (u_2)	3rd (u_3)	4th (u_4)	5th (u_5)	Total
Number caught	65	43	34	18	12	172
Expected number for $p = 0.33$	66.66	44.66	29.92	20.04	13.43	174.71
Chi-squared	—	—	—	—	—	1.06

The total catch, T, is calculated: $T = 65 + 43 + 34 + 18 + 12 = 172$.
Then

$$\sum_{i=1}^{k}(i-1)u_i = (1-1)65 + (2-1)43 + (3-1)34 + (4-1)18 + (5-1)12 \qquad (5.22)$$
$$= 213$$

is found.

Next the ratio R is determined, $R = 213/172 = 1.238$, and the estimate of the total population must be solved by numerical methods using a computer. However for $k = 3, 4, 5, 6$ or 7 these last steps may be circumvented by the use of Zippin's charts (Figs 5.3, 5.4).

Therefore in the present example for $k = 5$ and $R = 1.24$, the value of $(1 - q^k)$ is read off Figure 5.3 as 0.85, so that:

$$\hat{N} = \frac{172}{0.85} = 202 \qquad (5.23)$$

The standard error is given by:

$$SE(\hat{N}) = \sqrt{\frac{\hat{N}(\hat{N}-T)T}{T^2 - \hat{N}(\hat{N}-T)\left[\frac{(sp)^2}{(1-p)}\right]}} \qquad (5.24)$$

where the notation is as above and p is read from Figure 5.4.

In this example $SE(\hat{N}) = 14.46$. Therefore the 95% confidence limits of the estimate are approximately: $202 \pm 2 \times 14.46 = 202 \pm 28.9$.

If the lower confidence interval is less than the total number of captures, T, then T should be taken as the lower confidence interval.

It has been shown by Zippin (1956, 1958) that a comparatively large proportion of the population must be caught to obtain reasonably precise estimates. His conclusions are presented in Table 5.2, from which it may be seen that to obtain a coefficient of variation (CV = estimate/standard error × 100) of 30% more than half the animals would have to be removed from a population of less than 200. This requirement may make the method impractical for estimating populations of insects caught in pitfall traps; the proportion of the population captured is frequently too low.

After calculating estimates for N and p the expected number of captures should be calculated using equation (5.19) and the goodness of fit to the observed sequence of captures tested using a chi-squared test. The results of these calculations for the worked example are shown in Table 5.1 and indicate that a constant probability model can be accepted (χ^2 test statistic of 1.06, $k - 2 = 3$ degrees of freedom, probability of a value this large or larger can occur by random chance = 0.786). If a significant difference is found then the assumption that p is constant must be rejected.

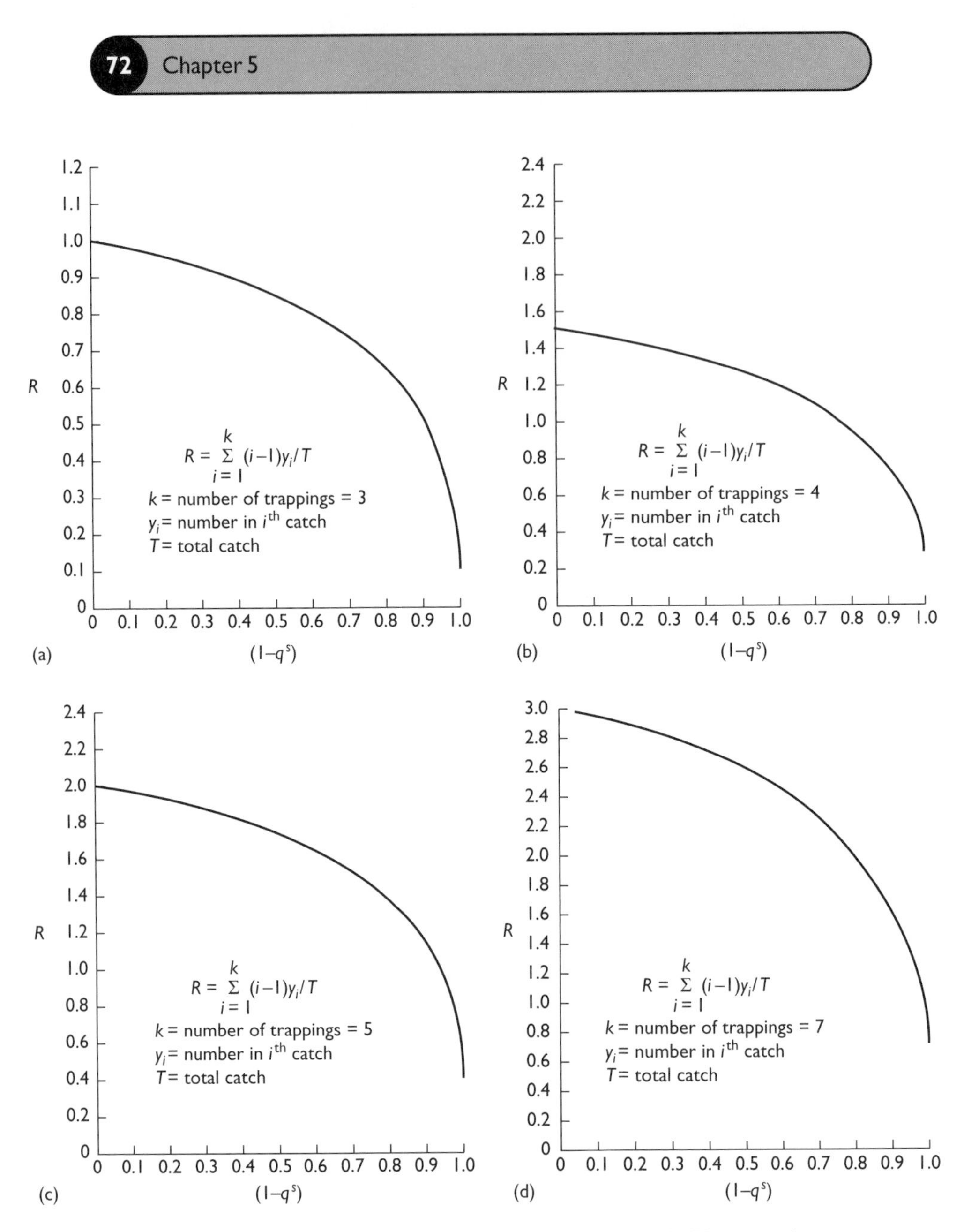

Figure 5.3 (a–d) Graphs for the estimation of $(1 - q^s)$ from the ratio R in removal trapping. (After Zippin 1956.)

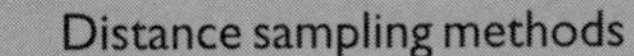

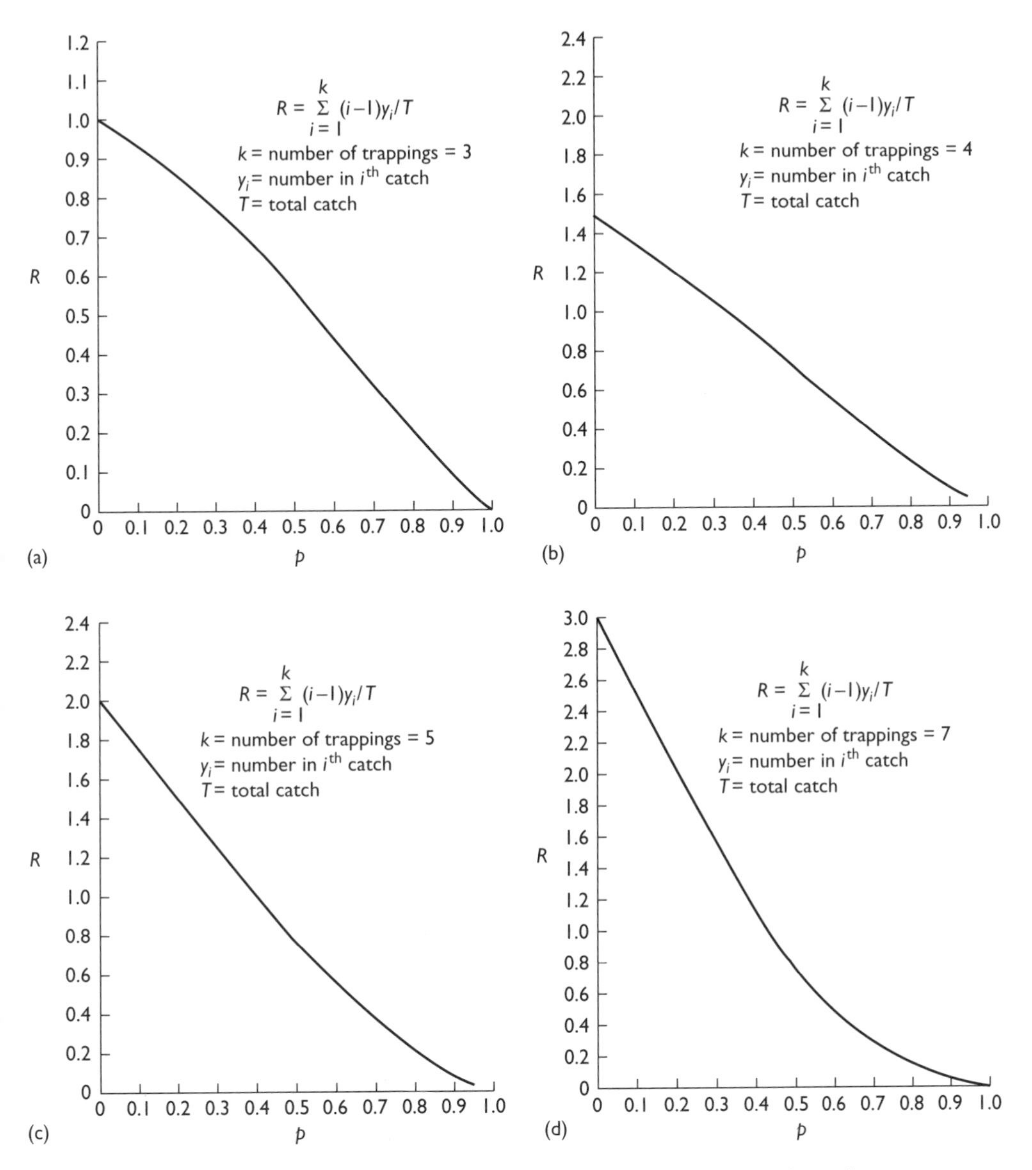

Figure 5.4 (a–d) Graphs for the estimation of p from the ratio R in removal trapping. (After Zippin 1956.)

Simplified procedures for when at most three sampling occasions have been undertaken are given in Seber (1982). For two samples a good approximation to the ML estimator is:

$$\hat{N} = \frac{u_1}{1 - \left(\frac{u_2}{u_1}\right)} \tag{5.25}$$

Table 5.2 Proportion of total population required to be trapped for specified coefficient of variation of N. (After Zippin 1956.)

N	Coefficient of variation			
	30%	20%	10%	5%
200	0.55	0.60	0.75	0.90
300	0.50	0.60	0.75	0.85
500	0.45	0.55	0.70	0.80
1,000	0.40	0.45	0.60	0.75
10,000	0.20	0.25	0.35	0.50
100,000	0.10	0.15	0.20	0.30

ea

Example applications

Electric fishing in streams

Quantitative electric fishing was undertaken on a small stream about 2 m wide with a maximum depth of about 70 cm. A 30 m reach was isolated by stop nets and electric fishing undertaken, working in the upstream direction so that any stunned fish tended to drift towards the samplers. All of the fish captured were identified to species and standard lengths were measured to the nearest millimeter. This procedure was repeated until the number of fish captured had declined over three successive passes. In this study only three passes were required ($k = 3$) and two species were sufficiently abundant to justify population estimation. The results obtained are shown in Table 5.3. At the end of the sampling all the captured fish were released alive.

Populations were estimated using Zippin's method and the population in a 100 m reach of the stream estimated (Table 5.4).

Animals in rock pools or hidden in the littoral zone

A wide variety of free-living animals in intertidal rook pools or hidden under stones or weed in the intertidal zone can have their populations estimated by removal sampling. Possibilities include shrimps, prawns, crabs, molluscs, and small fish. While some can be searched for by eye and removed by hand, other species in pools will need to be caught using a hand net. Delimit the search area and work over

Table 5.3 The number of trout and bullhead recorded over three successive passes along a 30 m stretch of stream.

Run	Bullhead	Brown trout
1st run (u_1)	40	13
2nd run (u_2)	11	6
3rd run (u_3)	7	4

(continued)

Example applications (continued)

Table 5.4 The estimated abundance of the two common fish in a small chalk stream in southern England.

Scientific name	Common name	Total number caught	Estimated population size (no. per 100 m of stream)
Cottus gobio	Bullhead	58	203.3
Salmo trutta	Brown trout	23	90.6

it systemically. An area that can be searched in about 15 minutes is probably optimal, as the area will need to be searched a number of times and the time before the tide rises will be limited. Stop the searches once the number caught has declined over three successive samples. The analysis is then undertaken as for the electric fishing example above.

Insects and snails on plants

Some large, slow-moving insects, such as some beetles (Coleoptera) and bugs (Hemiptera), can be picked by hand from leaves or flower heads. Another group that can be removed by hand are snails. Delimit an area of herbage that can be searched in 20 to 30 minutes and systematically move over the area collecting all of the target animals observed. Repeat this systematic search until the number obtained has declined over three successive samples. At the end of the study the animals can be returned alive to the habitat. The analysis is then undertaken as for the electric fishing example above.

For insects in grassland that are difficult to see or catch the area can be sampled using a suction sampler or by sweep netting. In this case, the samples will probably need to be killed in the field and subsequently sorted. It will therefore be difficult to know how many repeat samples should be undertaken, although a cursory examination will normally indicate if the number has fallen appreciably.

Chapter Six

Comparing the magnitude of populations – trapping and other relative abundance methods

Most relative methods often use simple, cheap apparatus and can produce impressive catches from situations where few animals would have been obtained using absolute methods. This makes them particularly appropriate for faunal surveys where plentiful data is needed and the costs of unit area sampling may be prohibitive. However, data collected by relative methods should be used and interpreted with caution. Comparisons of different species and different habitats are particularly problematic as the efficiency of sampling may be unknown and variable. Relative methods can be used for the following purposes:

1 To obtain a simple collection to gain some idea of what is present or for taxonomic study.

2 To obtain measures of availability, assuming the efficiency of the trap or search does not change, the raw data of catch per unit time or effort will provide a measure of availability. Over a period of time the changes in the species composition of the same trap in the same position may be used to indicate changes in the diversity of the fauna.

3 To obtain indices of absolute population. When the efficiency of the trap and the responsiveness of the animal to it can be regarded as constant and if the effects of factors that influence activity can be corrected for, then the resulting value is an index of the size of the population.

4 To estimate absolute population. It is sometimes possible to derive estimates of absolute population size from relative estimates by calibration with absolute estimates or by estimating the efficiency experimentally.

Factors affecting the size of the catch

There is no simple relationship between the size of the catch obtained by trapping and the size of the population. The majority or all of the following factors influence the relationship:

1 changes in actual numbers – population changes;

2 changes in the numbers of animals in a particular “phase” of the life cycle;

3 changes in activity following some change in the environment;

4 sexual or species-specific trap response;
5 changes in the efficiency of the traps or the searching method.

It is clear, therefore, that the estimation of absolute population by relative methods is difficult; what one is really estimating is the proportion of those members of the population that were vulnerable to the trap under the prevailing climatic conditions and the current level of trapping efficiency. The influence of factors 2–5 on these relative methods must be considered further below.

The influence of the developmental stage of the animal on the number caught

The susceptibility of an animal to capture or observation will alter with age if behavioral attributes or responses are age dependent. Many relative methods rely to some extent on the movements of the animal. For many terrestrial species, migratory movements usually occur early in adult life or between reproductive periods. The situation is often different for aquatic organisms such as fish where dispersal and migration often occur early in life. The reaction of animals to stimuli, important in many trapping techniques, also varies greatly over the life cycle.

Small mammals often vary in their vulnerability to trapping because of changes in their exploratory behavior and response to novel objects, which are in turn influenced by factors such as sex, age, and social status (e.g. Adler & Lambert 1997). Noting that feral house mouse populations have low recapture rates (0–20%) in live-trapping studies, Krebs *et al.* (1994) studied the movement of radio-collared mice. While low recapture rates during the breeding season were due to low trappability, during the nonbreeding period nomadic movements away from the study area were the cause.

Juvenile fish often vary in their vulnerability to light traps with age. In a study of North American freshwater fish, Gregory and Powles (1985) found common carp and bluntnose minnow entered traps almost exclusively as yolk-sac larvae, Iowa darter and pumpkinseed were taken only as yolk-sac and postlarvae, and yellow perch from 5 to 33 mm total length. There are no published studies that have examined individual differences in the vulnerability of adult fish to traps, although Bagenal (1972) suggested that the reproductive status of perch, *Perca fluviatilis*, might influence catch rate. Fish traps such as fyke nets and salmon butts are usually positioned along the banks of channels where they intercept fish undertaking seasonal, tidal or migratory movements and give catches biased with respect to age, sex, and reproductive condition.

Changes in the activity of the animal

The level of activity of possibly all animals follows a diurnal cycle, e.g. some insects fly by day, others at night, fish and mammals may forage at dawn and dusk, others by night or day. The weather almost always has a great effect on animal activity and the number trapped. The separation of changes in trap catch due to climate factors from those reflecting population change is a difficult problem that often cannot be solved. For insects it is frequently observed that catch and temperature are highly correlated. Occasionally it is possible to obtain a significant regression of catch size on temperature for a single species, particularly for groups such as fish that increase their swimming activity with temperature. Many insects, crustaceans, fish, reptiles and other cold-blooded animals have a threshold temperature below which they do not become active. Insects may have upper and lower temperature thresholds for flight. If activity is known to occur between thresholds, then once these have been determined fluctuations in numbers between the lower and upper thresholds may be considered to be independent of temperature. Other climatic variables of great importance are wind and rain. Small mammals also show changes of activity with respect to variables such as temperature, cloud cover and rain which change the number trapped. Although subjected to little quantitative study, changes in physical factors such as water temperature, flow, light, tide or oxygen concentration greatly alter the activity of fish and the likelihood of capture.

Differences in the response between species, sexes, and individuals

Our ability to observe or catch different species varies greatly and this must be taken into account when choosing methods for faunal surveys. Small mammal species differ in their response to novel objects (Myllymaki *et al.* 1971) and those with a higher aversion are caught less frequently. For example, Feldhamer and Maycroft (1992) found the mean number of captures per individual for golden mice, *Ochrotomys nuttalli*, were significantly less than those for white-footed mice, *Peromyscus leucopus*. Fish species vary greatly in their trappability. Pelagic (open-water) species may avoid structures and thus never enter traps while some species that live within floating meadows or weed beds may be so adept at navigating maze-like environments that they can escape. Species also differ greatly in their ability to detect and avoid gill nets, which are particularly ineffective against electric fish and other elongate, nonspiny forms that do not use sight for navigation. Individual fish differ in their willingness to explore new objects and thus become trapped. There are reports that the presence of fish alarm substance which is released from injured skin will deter fish such as fathead minnows, *Pimephales promelas*, from entering traps (Mathis & Smith 1992). While in this case the existence of pheromones has been questioned (Magurran *et al.* 1996; Henderson *et al.* 1997), the general principle that traps should be kept clean and give no signal of their prior use is sound practice.

In many groups significantly more of one sex than the other are caught in traps. For example, mirid bug males make up the majority of light-trap catches and in some species of moth the females will never be caught as they do not fly. The positioning of traps can influence the sex ratio bias. An interesting case of a difference between sexes is the large number of male relative to female tsetse flies usually taken on "fly rounds" (Glasgow & Duffy 1961). The biological interpretation of this seems to be that newly emerged female flies usually feed on moving prey and the early pairing, desirable in this species, is achieved by numbers of males following moving bait. For mammals, sex differences are also common; mature males are often caught more easily than females, but the reverse can be the case. For example, Adler and Lambert (1997) found that adult females of the Central American spiny rat, *Proechimys semispinosus*, were more trappable than males.

Individual differences in capture rates have been related to genetics. Gerard *et al.* (1994) found that heterozygous house mice, *Mus domesticus*, obtained by the crossing of two chromosomal strains had a lower trappability than the homozygous wildtypes. Diseases and parasites can also affect the trap response and should always be considered as a potential source of bias in a survey of infection rates as it is frequently easier to trap sick animals. Webster *et al.* (1994) found that infection with *Toxoplasma gondii* changed avoidance behavior in wild brown rats, *Rattus norvegicus*, resulting in the more frequent capture of infected individuals. Similar factors have been suspected to operate in fish populations.

The efficiency of a trapping or searching method

The efficiency of a method of population estimation is the percentage of the animals actually present that are recorded. The efficiency of a searching method depends both on the skill of the observer and the habitat; e.g. any observer is likely to see tiger beetles (Cicindellidae) far more easily on lacustrine mud flats than on grass-covered downlands. The weather affects the efficiency of many traps; e.g. sticky, water and flight interception traps catch insects that are carried in to or on to them by the wind and efficiency varies with wind speed. Light traps are also affected by wind and generally catch fewer specimens of most insects on nights when the moon is full. Other variables noted by Martin *et al.* (1994) to influence sweep net and sticky trap sampling efficiency for black flies, *Simulium* sp., were wind speed, light, temperature, saturation deficit, and time of day. The effectiveness of a given bait may even vary from habitat to habitat. Fish traps and gill nets vary in efficiency with changes in water turbidity, turbulence, light, flow, temperature, and depth.

Example applications

A selection of relative sampling techniques are described below. Methods for trapping small mammals are described in Chapter 3.

Fish traps in ponds, rivers, and docks

Man has trapped fish for thousands of years and a great many methods are still currently used. One of the most familiar is the lobster pot, which is used throughout the world. It is useful to study local methods, which are designed to perform under the specific conditions of water and animal behavior. Most fish traps work on the principle that it is easier for a fish to enter than leave and thus the catch need not be all the fish that have entered. Commonly used types of fish trap, illustrated in Figure 6.1, include fish weirs, fyke nets, pot traps, salmon putts, and minnow traps. All fish traps are highly selective. Aquatic traps can be used with an attractant such as light (see page 77) or a bait, although this is rarely necessary.

The minnow trap is simple to construct and is a particularly effective small fish trap that can be used to collect animals for a mark–recapture experiment in a small pond. It will catch a wide range of small fish. It comprises a funnel inserted into a bottle and can often be quickly and cheaply constructed from a plastic drink bottle. The mouth of the bottle is simply chopped off and pushed onto the end in the reverse direction (Fig. 6.1a). Alternatively, these traps may be made by attaching plastic filter funnels to the mouth of a screw-top jar. Traps for larger fish can be quickly constructed using plastic or metal mesh (Fig. 6.1c). A type of trap that can be bought commercially and is particularly effective in estuaries and along the edge of large rivers is a fyke net (Fig. 6.1d). Traditionally these nets are used for catching eels, but they are effective for many species. In the Amazon they are remarkably effective for catching electric fish. The effectiveness of fish traps depends on their positioning and to some extent the best position should be discovered by trial and error. However, it should be remembered that fish do not need to enter traps, so they should be positioned in areas that are attractive for fish to enter. For example, minnow traps may be more effective if placed in areas where there is cover, such as amongst vegetation along the bank.

Gill nets (Fig. 6.2)

Gill nets are designed to work by catching swimming fish that attempt to pass through the net, but when they find the mesh is too small, they cannot reverse out. This is generally because their gill covers catch on the net, but, in the case of some catfish, they are entangled by their spiny pectoral fins. The size range of fish caught is related to mesh size and it is common practice during general fisheries surveys to lay banks of nets with a range of mesh sizes. The efficiency is related to the ability of fish to detect the net. Fish with good vision are caught more often when light levels are low. However, this does not imply that they are always more effective in the dark as they will only catch swimming fish and many fish are inactive at night (while other species become active). Monofilament nets are more difficult than multifilament nets for fish to detect and thus tend to be more efficient. However, they are more easily damaged by predators such as crabs or caimans, which are attracted by the catch.

Gill nets can kill large numbers of fish and other aquatic life and their use is often restricted. You should ensure that they are legal to use in your proposed study area.

(*continued*)

ea Example applications (*continued*)

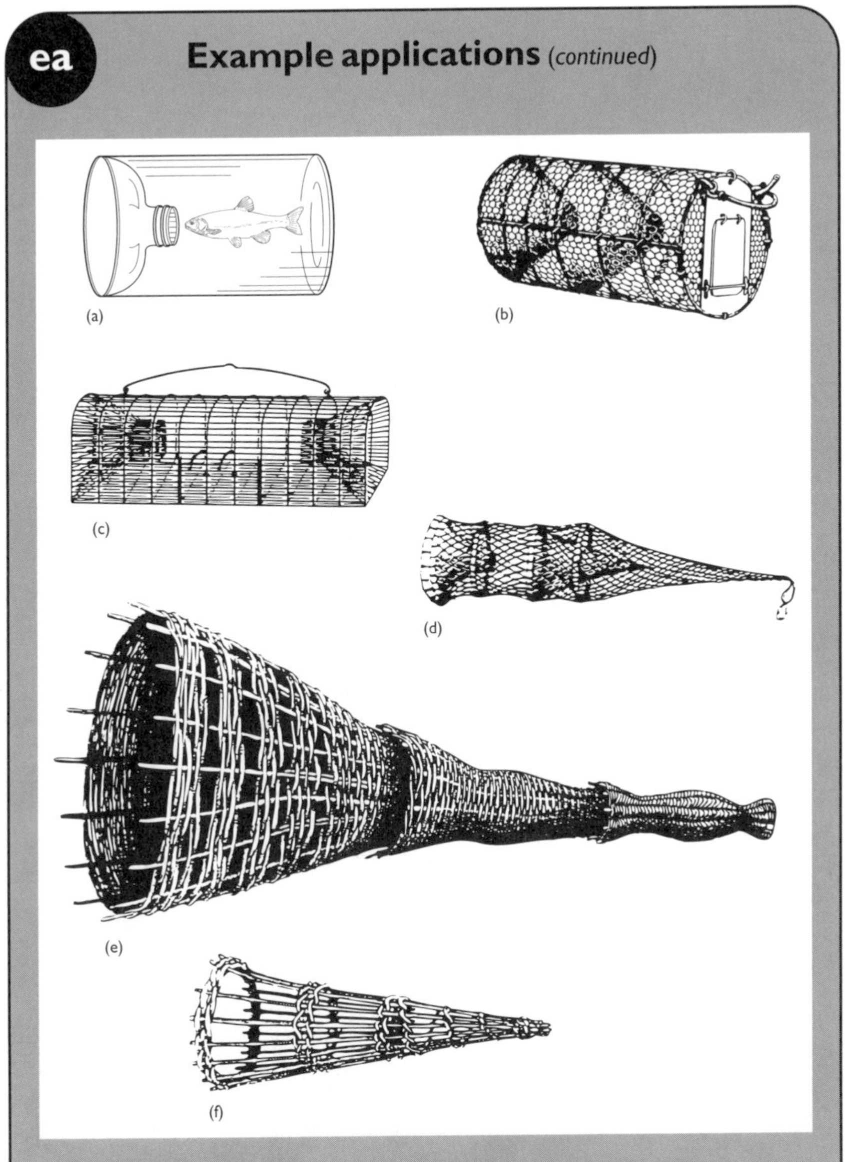

Figure 6.1 Fish traps. (a) Minnow trap made from plastic drink bottle with the neck cut off and inverted. (b) Eel grig. (c) Wire mesh double entry trap. (d) Dutch fyke net. (e) Putt – traditional Severn estuary trap made from woven willow. (f) Putcher – traditional River Severn trap.

Fish abundance is normally expressed as catch per unit effort where effort is given in units of net length × time. This expression might suggest that the catch in a net increases linearly with time, which is not true. Gill nets are probably best used for comparing the abundance and diversity of fish in different localities within a lake or coastal

(*continued on page 82*)

Example applications (*continued*)

Figure 6.2 A gill net fixed to sample fish in the intertidal zone. The fish swim into the net and cannot escape because their gill covers do not allow them to reverse out.

waters. Unfortunately, the fish are usually killed or badly harmed and many cannot be returned alive.

Stream kick net sampling (see Fig. 3.11)

This method is suitable for obtaining samples of bottom-living invertebrates from streams. A hand net with a flat top edge is placed touching and perpendicular to the substrate and facing into the flow. The area in front of the net is disturbed with the foot by repeated "kicking" across the face of the net. Frost *et al.* (1971) found that the first kick disturbed about 60% of the fauna yielded by 10 kicks. A more precise delimitation of the area is used in the Surber sampler (see Fig. 3.12). The catch of animals displaced from the bottom is washed from the net into a tray where they can either be immediately sorted and inspected or transferred to a container and treated with formalin for future sorting and identification.

(*continued*)

ea

Example applications *(continued)*

Kick net sampling is particularly appropriate for the comparison of aquatic life in different reaches of a stream or between streams. It is often used in conjuction with methods such as the BMWP scoring system (see page 126).

Hand netting in ponds and stream for invertebrates and small fish

Anyone who as a child did not pursue pond life with a hand net has been deprived of a great pleasure. The method is also useful to ecologists and is often the only means available for the capture of small fish and insects in weedy ponds. An example of the use of hand nets is the study of the fauna of submerged leaf-litter. In shallow waters, it is the most practical way to lift a portion of a substrate. Henderson and Walker (1986) used shallow hand nets of 0.1 m^2 area to quantitatively sample the insect and fish community living within Amazonian stream litter banks. The net was rapidly thrust into the side of a submerged litter bank and raised rapidly vertically to the surface. The catch was then washed free of the leaves in the open water of the stream, transferred to small jars, and preserved in 70% alcohol.

Pooting and visual searching in grazed grassland

The aspirator or pooter (Fig. 6.3) is convenient for rapidly collecting small insects in fixed time estimations. When using a pooter care should be taken to ensure the operator cannot accidentally inhale damaging particles. Evans (1975) describes two pooter designs that work on the "Venturi" principle, the operator blowing. The one shown in Figure 6.3 is emptied by carefully removing the top cork and the gauze filter; this operation and the realignment of the tubes before reuse must be carried out carefully. Sampling efficiency tends to vary with the weather and time of day. Pooting is best undertaken on closely mown or grazed grassland. The small insect fauna of different areas, perhaps slopes facing north and south, can be compared using collections taken over a set timed period in each locality.

Interception traps

There are many types of interception trap that can be used to sample flying insects. One of the simplest and least expensive is a window trap. It is useful for sampling flying beetles (Coleoptera) and other insects that fall upon hitting an obstacle. A window trap is basically a large sheet of glass or clear plastic held vertically, with a trough containing water with a wetting agent and a little preservative below it (Fig. 6.4). The traps should be placed across possible flight paths of insects and the trapped insects collected daily. This type of trap can be used to compare insect faunas from different localities.

A larger and more costly trap particularly suited for sampling flies (Diptera) is the Malaise trap (Figs 6.5, 6.6). It is basically an open-fronted tent and presumably the idea came from the observation of trapped insects while camping. Townes (1962) gives full instructions for the construction of his model, which is durable and traps insects from all directions. Malaise traps should be placed across "flight paths" such as woodland paths, but in windy situations they cannot be used.

(continued on page 84)

Example applications (*continued*)

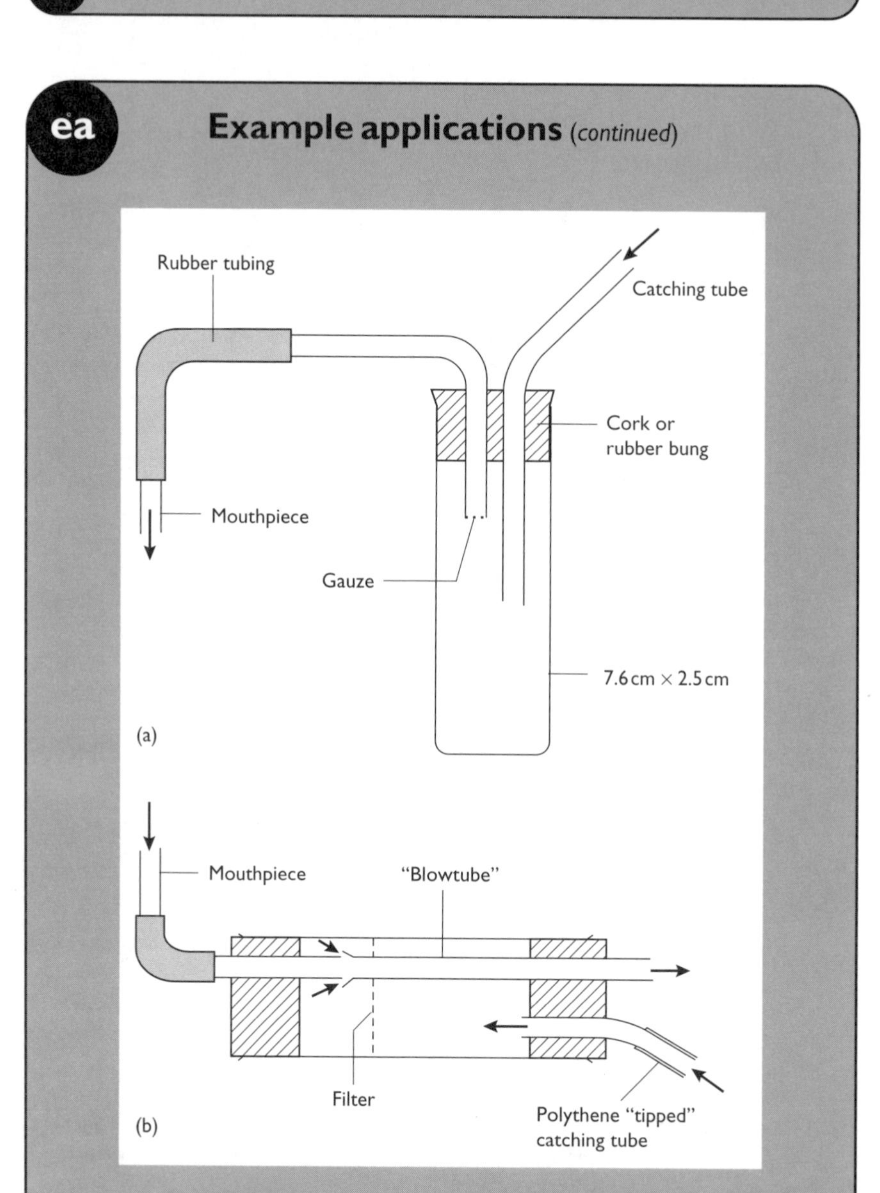

Figure 6.3 The aspirator or pooter: (a) a simple model; (b) a blow model (after Evans 1975) – mouthpiece is 9 mm diameter and the blowtube 11 mm external diameter.

Light trapping for insects, particularly moths

Light traps using ultraviolet light are probably the most widely used insect traps. The exact mechanisms that lead to insect captures by a light trap are far from clear and one should be aware that the catch may not accurately reflect either the species complement of the habitat or the relative abundance of species. Insects can be both attracted and repelled by light. The catch of night-flying insects varies with the phase of the moon, temperature, wind, humidity, and other environmental variables. Light traps

(*continued*)

ea Example applications *(continued)*

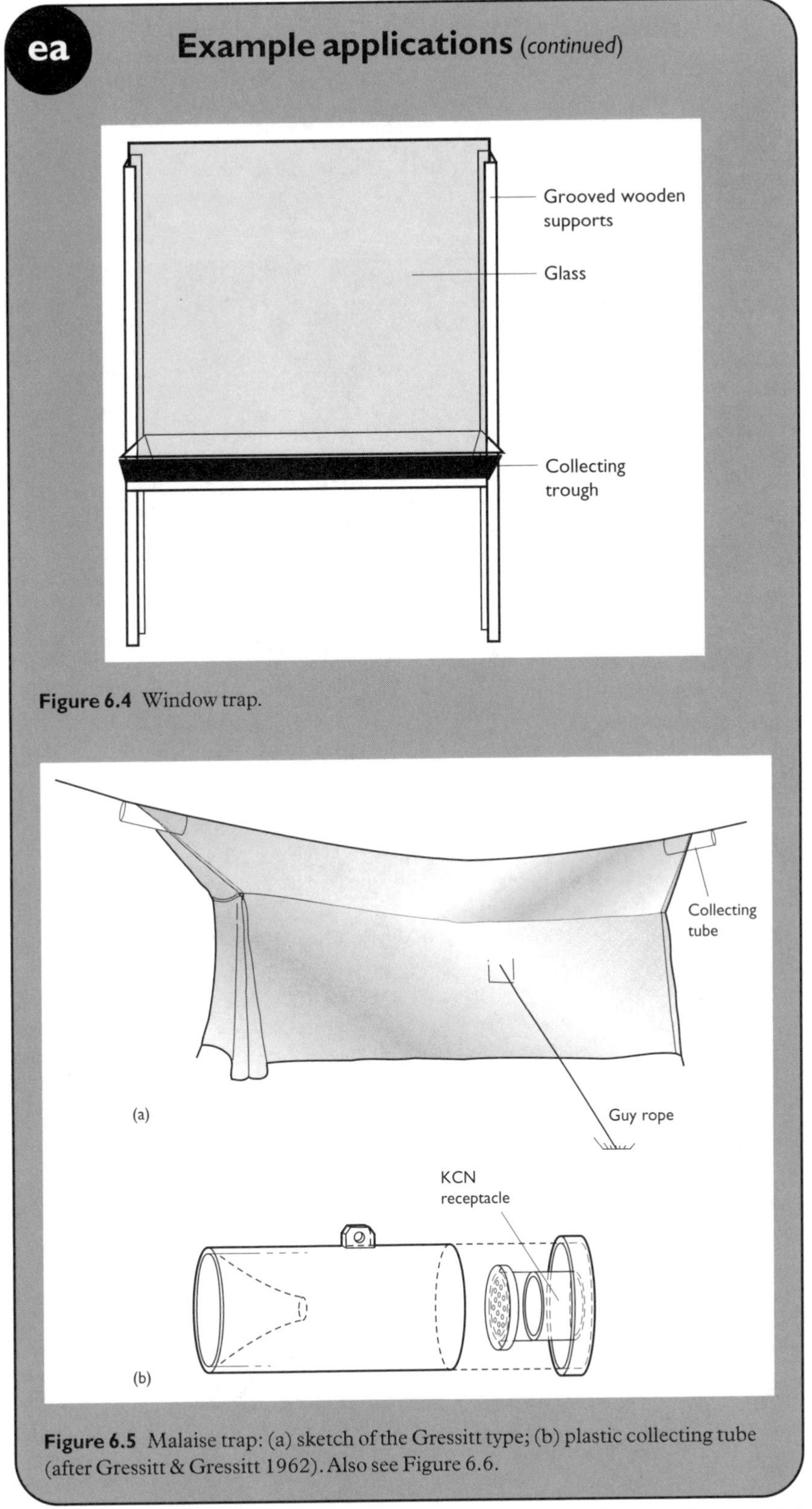

Figure 6.4 Window trap.

Figure 6.5 Malaise trap: (a) sketch of the Gressitt type; (b) plastic collecting tube (after Gressitt & Gressitt 1962). Also see Figure 6.6.

(continued on page 86)

ea Example applications (*continued*)

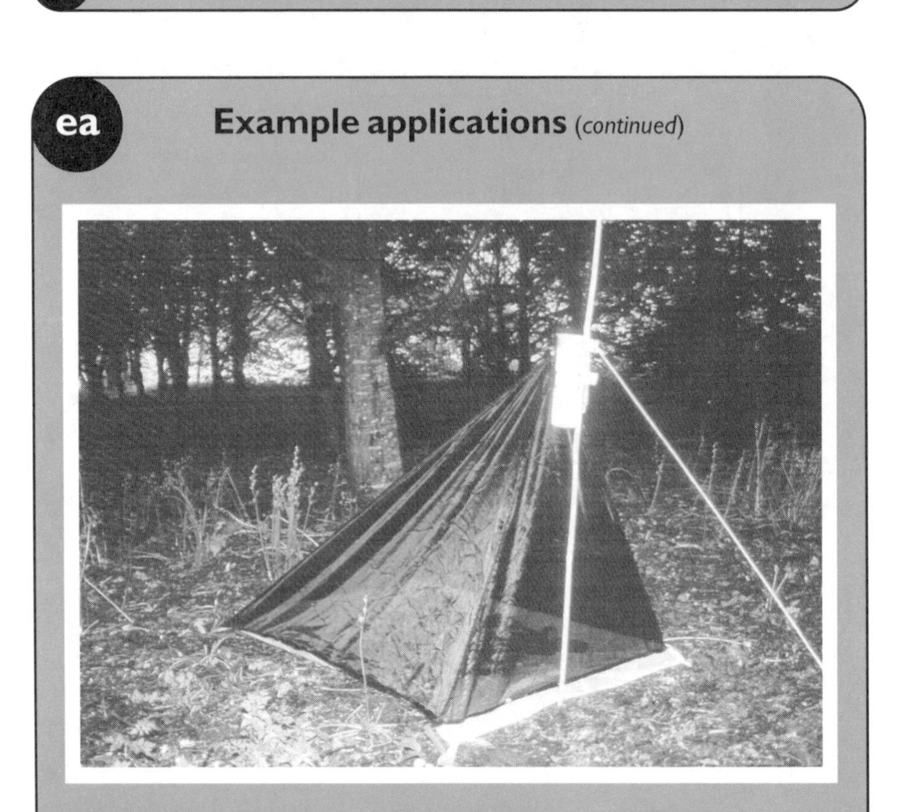

Figure 6.6 A Malaise trap.

have been found useful in survey work, both for particular species and general studies of a taxonomic group of insects in a region.

Many light traps retain the insects alive; the Robinson trap retains the insects in the drum of the trap, but small moths may leave the trap in the morning. A wide variety of trap designs are available. Well-known designs and their attributes are summarized below.

- The Rothamsted trap (Fig. 6.7a). This trap takes moderate numbers of most groups, including nematocerous flies and the larger moths; it would seem to be less selective than most other light traps and is useful in studies on the diversity of restricted groups.
- The Robinson trap (Fig. 6.7b). This was the first trap using ultraviolet light and was designed by Robinson and Robinson (1950) to make maximum catches of the larger Lepidoptera. It is therefore without a roof and although a celluloid cone protects the bulb and there is a drainage hole in the bottom, it cannot be used in all weathers. The drum is partly filled with egg boxes or similar pieces of cardboard in which the insects can shelter. The catch may be retained alive. Large numbers of fast-flying nocturnal insects (e.g. Noctuidae, Sphingidae, Corixidae, Scarabaeidae) are trapped.
- The Pennsylvanian and Texas traps (Fig. 6.7c). These traps are similar, consisting of a central fluorescent tube surrounded by four baffles; below the trap is a metal funnel and a collecting jar. Fluorescent tubes may be run for many hours from batteries. The Minnesota and Monks Wood light traps are basically similar to the Pennsylvanian model, but the roof is cone shaped rather than flat and this appears to reduce the catch.

(*continued*)

ea

Example applications (*continued*)

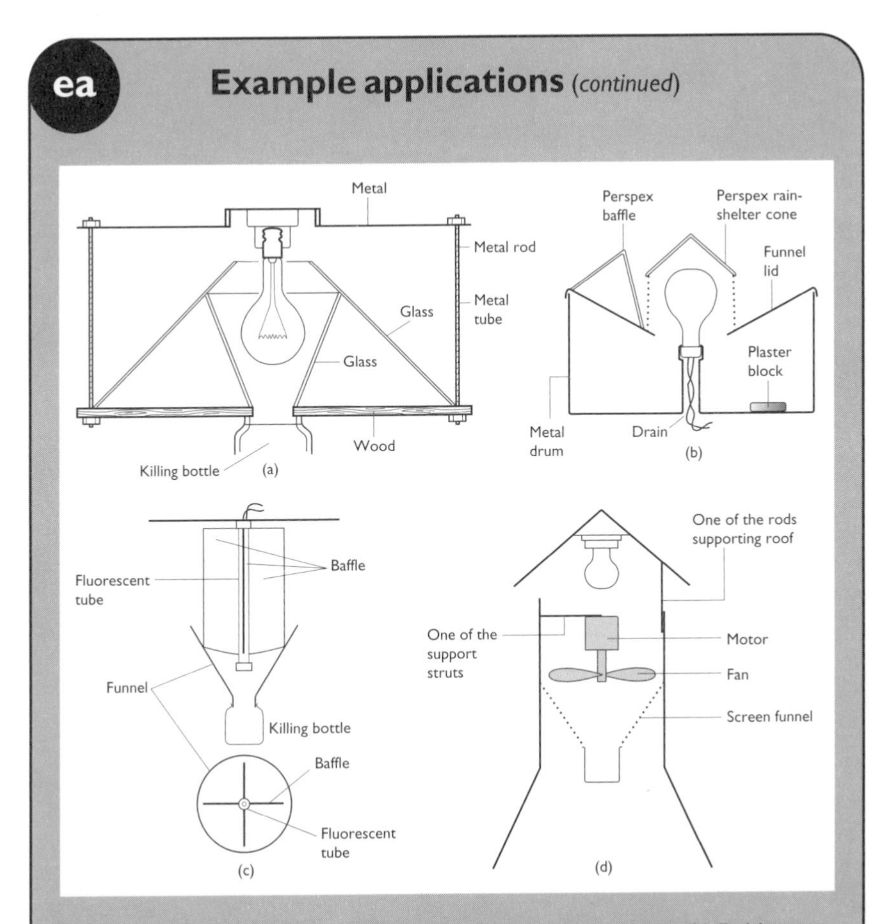

Figure 6.7 Light traps: (a) Rothamsted trap (after Williams 1948); (b) Robinson type trap; (c) Pennsylvanian trap (sectional and plan views); (d) New Jersey trap. Also see Figure 6.8.

• The New Jersey trap (Fig. 6.7d). This is primarily a trap for sampling mosquitoes; it combines light and suction – the lamp causes the insects to come into the vicinity of the trap and they are drawn in by the suction of the fan. The trap is therefore particularly useful for weak flyers that may fail to be caught by the conventional light trap. Fitted with an ultraviolet lamp the New Jersey trap has been used to collect other groups of small insects.

Light trapping is a useful method to compare the diversity of moth or bug (Hemiptera) faunas of different habitats. The traps should be run for a number of nights and the catch identified to species and counted each morning (Fig. 6.8). A careful note should also be taken of meteorological conditions and their effect on the catch over the study period investigated. The diversity of the catch in, for example, different types of woodland can be compared using diversity and similarity indices, and multivariate methods (see Chapter 10).

(*continued on page 88*)

ea Example applications (*continued*)

Figure 6.8 A portable light trap. The light source is a 12 V fluorescent tube that will run all night on a small car battery. The light is enclosed within three clear plastic veins that insects will hit when circling the light, knocking them via a cone into the holding box below. This type of trap will catch hundreds of insects per night.

Pitfall trapping for ground-living insects and spiders (Fig. 6.9)
Like the lobster-pot, the pitfall trap was an adaptation by the ecologist of the technique of the hunter; basically it consists of a smooth-sided hole into which the hunted animal may fall but cannot escape. Often a glass, plastic or metal container is sunk into the soil so that the mouth is level with the soil surface. Pitfalls can be used to capture arthropods, amphibians, snakes, and mammals. Pitfalls have many advantages: they are cheap (empty food or drink containers may be used), they are easy and quick to operate, and a grid of traps can provide an impressive set of data. However, catch size is influenced

(*continued*)

ea Example applications (*continued*)

Figure 6.9 A pitfall trap for insects and other terrestrial arthropods such as spiders. The trap is a disposable cup set into the ground. It is protected from rain by a simple roof made of corrugated plastic.

by a wide range of factors, apart from population size and, as shown by Topping and Sunderland (1992) in a study of the spider fauna of winter wheat, the catch may not reflect changes in abundance, the relative abundance between species, nor even the sex ratios of species. Pitfalls can be arranged in trapping webs or grids. For small mammal studies, an advantage of pitfalls over snap or live traps such as the Longworth is that they can produce multiple captures, although a captured animal may influence the probability of further captures. Pitfalls have been used extensively for studies on surface dwellers such as spiders, centipedes, millipedes, ants, beetles, especially Carabidae, and even crabs.

Baits can be used in pitfall traps and baited traps have been found useful for collecting beetles. However, the effects of baits are variable and unbaited taps are generally to be preferred for general collections. A possible arrangement for a baited trap suitable for dung or carrion feeding insects is shown in Figure 6.10.

Pitfalls are particularly appropriate for ground-living beetles, ants and spiders where the aim is to compare the diversity between habitats (see Chapter 10 for the estimation of diversity indices). A suitable container is a disposable drinking cup that must be placed in the soil so that the lip is level with the surface of the soil (Fig. 6.9). Retention of a captured animal depends on the walls of the trap being smooth and clean. Winged species may escape by flight or they may be eaten by larger species. This can

(*continued on page 90*)

Example applications (*continued*)

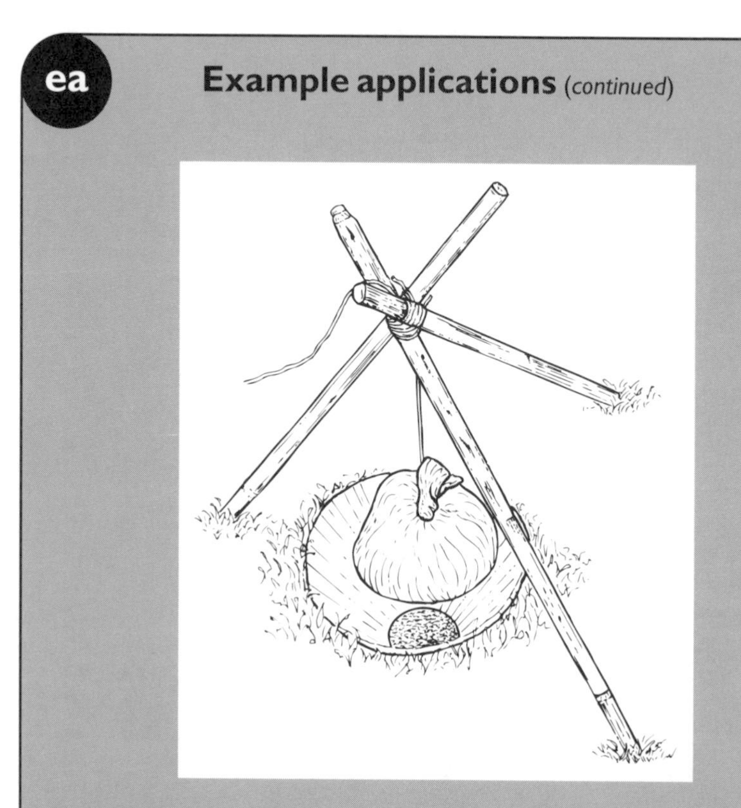

Figure 6.10 A possible arrangement for a baited trap for carrion and dung feeding insects.

be avoided by adding water with a small amount of detergent as a wetting agent to each trap. Preservatives such as formalin or alcohol have sometimes been used in traps, but these may affect the catch. Ethylene-glycol antifreeze has been found to work better for the Carabidae. It is important to ensure that the trap will not be flooded by rain or surface water. Small roofs should be fitted above the cups if rain is possible.

The traps can be set in a wide variety of habitats including saltmarsh, sand dune, grassland, and forest. They should normally be set for at least 24 hours between collections, which should be undertaken over a number of days. If time allows, pitfalls charged with ethylene-glycol can be left in place for a month or more.

If the aim is to return animals alive from pitfalls then they must be protected from flooding and will need to be fitted with drainage holes in addition to a roof.

Sweep netting in grassland or among annual crops for insects and spiders (Figs 6.11, 6.12)

The sweep net is perhaps the most widely used piece of equipment for sampling insects from vegetation; its advantages are simplicity, speed, and high return for little cost. Further, it will collect sparsely dispersed species. However, only those individuals on the top of the vegetation that do not fall off or fly away on the approach of the collector are caught. A sweep net cannot be used on very short vegetation. If plants are taller than about 30 cm, further increases in height mean that the net will be

(*continued*)

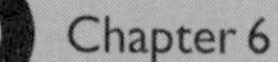

ea Example applications *(continued)*

Figure 6.11 A selection of nets used for terrestrial insect and arthropod sampling. From left to right, a small collapsible pocket net, a standard sweep net for sampling small animals in vegetation, a kite net for butterflies and moths, and a large net for flying insects.

sampling progressively smaller proportions of any insect population whose vertical distribution is more or less random.

Changes in efficiency may be due to:

- changes in the habitat;
- changes in species composition;
- changes in the vertical distribution of the species being studied;
- variation in the weather conditions;
- the influence of the diel cycle on vertical movements.

(continued on page 92)

ea Example applications *(continued)*

Figure 6.12 A sweep net in use collecting bugs (Heteroptera) from grass.

The bags of sweep nets are usually made of linen, thick cotton, or some synthetic fiber; the mouth is most often round, but a D-shaped mouth is useful for collecting from short vegetation, especially young crops. Usually, the more rapidly the net is moved through the vegetation the larger the catch.

Sweep netting can be used to estimate the diversity and abundance of grassland spiders. Luczak and Wierzbowska (1959) considered that undertaking 25 sweeps in 10 separate areas would give a sample of adequate size.

Water and sticky traps for small insects such as aphids

Water traps are simply glass, plastic or metal bowls or trays filled with water to which a small quantity of detergent and a preservative (usually a little formalin) have been

(continued)

ea Example applications *(continued)*

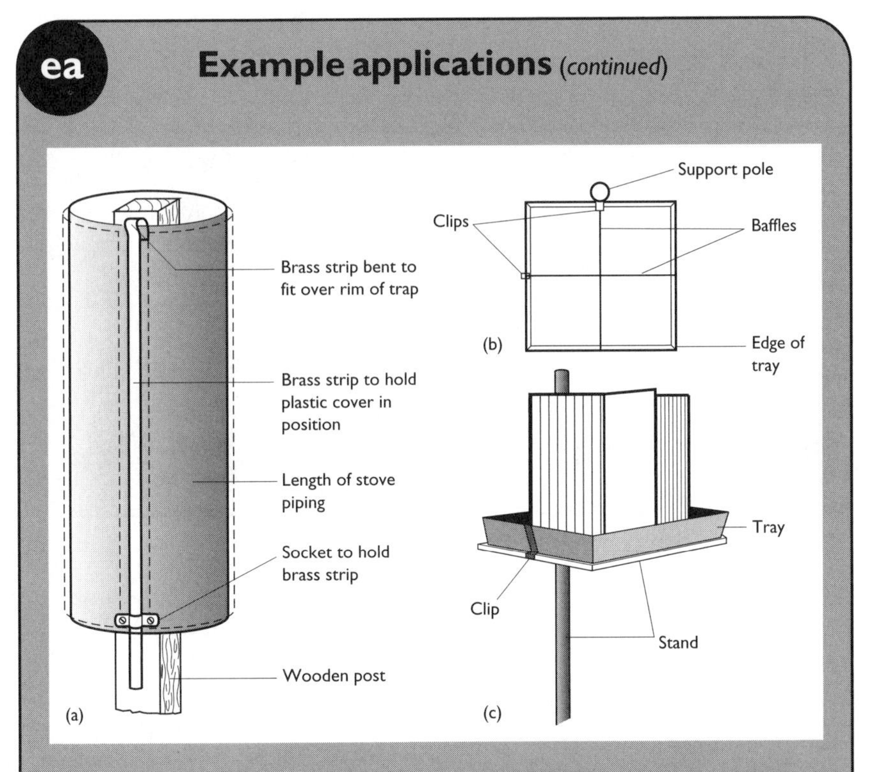

Figure 6.13 (a) Cylindrical sticky trap (after Broadbent *et al.* 1948). (b,c) Plan and isometric view of a water trap with baffle (design reported by Coon & Rinicks 1962).

added. Omission of the detergent will more than halve the catch. The traps may be transparent or painted various colors and placed at any height. Yellow has long been recognized as one of the most effective colors. A variety of colors was tested by Harper and Story (1962) for the sugar beet fly, *Tetanops myopaeformis*. The total numbers taken in the different colored traps were: yellow, 330; white, 264; black, 202; red, 107; blue, 64; green, 53. The "catching power" of a water trap can be increased by standing two upright baffles at right angles to each other to form a cross in the tray (Fig. 6.13). These four divisions created allow the separation of the insects according to the quadrat in which it has been captured; this may be related to the direction of flight at the time of capture.

Sticky traps work by retaining animals that settle or impact the adhesive surface. A variety of adhesives may be used; those resins and greases developed for trapping moths ascending fruit trees have proved particularly useful; castor oil may be satisfactory for minute insects. One of the most commonly used sticky materials is polyisobutylene and instructions for the freeing of insects from this matrix are given by Murphy (1985). The insects are separated from most fruit-tree banding resins by warming and then scraping the resin plus insects into an organic solvent like trichlorethylene or hot paraffin from which they may be filtered. The separation from the greases is easier; a mixture of benzene and isopropyl alcohol rapidly dissolves the

(continued on page 94)

ea Example applications *(continued)*

adhesive. Thus when possible, a grease will be used in preference to a resin, but only weak insects (mosquitoes, mites and aphids) will be trapped by a grease.

There are many sticky trap designs, but most traps comprise large screens or small cylinders, boxes, or plates. Cylindrical plastic traps are commonly used for small insects. Large screens consisting of a wooden lattice or series of boards have been used to measure movement at a range of heights of aphids and beetles. Glass plates about 20 cm square, with adhesive spread on the upper surface and underpainted yellow, are useful for small insects. Glass has the advantage that the adhesive is easily scraped off with a knife or the whole plate can be immersed in a solvent.

Water and sticky traps can be used to compare the activity of insects under different climatic conditions. They can also be used to compare the diversity of different habitats.

Chapter Seven

Using signs and products as population indices

Comparative surveys of the abundance of some animal groups can be undertaken by surveying signs or products such as footprints, feces, nests, burrows, or cast skins. Measures of the size of populations based on the magnitude of their products or effects are often referred to as population indices. The relationship between these indices and the actual population size varies from equivalence when, for example, the number of shed exuviae of an insect are counted, to no more than an approximate correlation, when the index is obtained from general measures of damage. Below are given just a few examples of the types of signs that might be used in a variety of habitats.

Leaf damage on trees or shrubs

In some cases the damage done by a single insect can be used to estimate the number passing through a particular stage of the life cycle. For example, if each leaf miner leaves a recognizable unique mark on a leaf, these can be counted and after all the animals have passed this stage, a measure of the total number passing through that stage calculated. As another example, some stem-borers cause the growing shoot to die; when multiple invasion is sufficiently rare to be overlooked, estimates of these "dead hearts" may be taken as equivalent to the total number of larvae invading the crop.

The actual amount of leaf damage caused by insect herbivores may be assessed in terms of dry weight reduction or leaf area destroyed. The area of the leaf may be measured electronically, by using a flatbed scanner with a computer, or by photocopying and weighing. The area of leaf blotching due to the removal of cell contents (by leafhoppers and mirids) may be estimated by color scanning followed by image analysis in which the area occupied by different colors is calculated by a computer. The number of pixels with each color can be calculated using many standard photomanipulation packages.

It is often easier to measure some index of the amount of plant destroyed rather than the actual quantity. Nuckols and Connor (1995) visually examined each sampled leaf for damage caused by chewing herbivores, skeletonizers, sap feeders, leaf miners, and gall formers, and scored the effect in terms of the percentage of the leaf area affected. Estimates of the quantity of leaf consumed may be confused by the subsequent enlargement of the hole by the growth of the leaf. Lowman (1987) showed for a range of tropical forest trees that leaf holes expand proportionately with leaf growth.

Frass as a measure of tree-living insects

The feces of insects are generally referred to as frass and frass-drop. The number of frass pellets falling to the ground was first used as an index of both population and insect damage by a number of forest entomologists in Germany.

The falling frass is collected in cloth or wooden trays or funnels under the trees (Fig. 7.1). In order for such collections to be of maximum value for population estimation one should be able to identify the species and the developmental stage. Information is also required on the quantity of frass produced per individual per unit of time and the proportion of this that falls to the ground and is collected. If a study of an individual species is planned then the frass must be identifiable to species. You might be surprised to know that in some countries, keys for the identification of frass pellets have been prepared (Morris 1942; Weiss & Boyd 1950, 1952; Hodson & Brooks 1956; Solomon 1977).

A greater proportion of the frass will be retained, on the foliage, in calm weather than under more windy conditions, and where the young larvae produce webbing, the frass will tend to be caught up with this and its fall delayed. Perhaps the simplest example of the potential for measuring frass-drop can be obtained from a laboratory study of a large insect such as a stick insect (*Phasmida*).

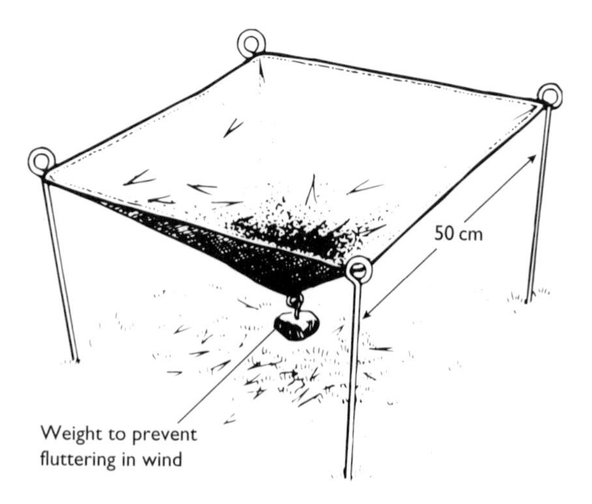

Figure 7.1 A simple cloth frass collector. (After Tinbergen 1960).

Discarded pupal cases around ponds and exuviae from trees

The larval or pupal exuviae of insects with aquatic larval stages are often left in conspicuous positions around the edges of water bodies, and where it is possible to gather these they will provide a measure of range, species diversity, emergence rate, and absolute population of newly emerged adults. The method is most easily applied to large insects such as dragonflies. Surveys of chironomid communities in streams and rivers have frequently been undertaken by collecting floating pupal exuviae.

Paramonov (1959) drew attention to the possibility of obtaining an index of the population of arboreal insects from their exuviae, more particularly from the head capsules of lepidopterous larvae that may be collected in the same way as frass (see above).

Assessing density by counting webs in grassland or heath

Populations of web-building spiders can be estimated by counting the number of webs. The visibility of the webs may be increased by dusting with lycopodium powder or by spraying with a fine mist of water. Indices of the populations of colonial nest-building caterpillars have been obtained by counting the number of nests, rather than the caterpillars. Some mites also produce webbing and this may be used as an index.

Using casts to estimate the activity of worms on land or on intertidal flats

Earthworms and marine polychaetes such as the lugworm form characteristic casts on the surface of the soil or sand that can be counted. A measure of the density (or activity) of the worms per unit area can be measured by counting the worm casts per unit area. The abundance of worm casts can be estimated by counting the number of casts within a randomly placed quadrat (see page 24). Alternatively, it may be better to run a line transect along the study area and count the number of casts in a quadrat placed at regular intervals.

The marks and signs made by fish

Mullet are marine and estuarine fish that feed on diatoms on the surface of intertidal mud flats. At low tide the spots where the mullet have been grazing the diatoms off the surface of the mud can be seen because the fish leave characteristic paired lines on the surface of the mud. Counting the number of marks per unit area can be used to compare the feeding activity of mullet in different areas.

Other fish can also leave feeding signs. For example, in the Amazon Basin the activity of piranha can be measured by the frequency of bite marks on the tails of other fish species. In some temperate rivers the presence of lamprey can be measured by the presence of feeding scars on other fish.

Gravel-spawning fish such as trout and salmon can be quantified by counting redds, the characteristic areas of disturbed gravel where they have buried their eggs. These are often placed in shallow water and can be counted by eye while walking a set distance of stream. The nests of sunfish can also be counted in some waters.

Terrestrial vertebrate signs

Bird, reptile and mammal abundance can be estimated from counting nests or burrows. The main problem is to ensure that the structure is occupied. Nest counting is particularly appropriate for seabirds and other colonial breeders that synchronize their reproduction. The census techniques to be used will vary with the habitat and for seabirds are discussed by Walsh *et al.* (1995).

Bird and mammal numbers may be estimated by counting feces. For many mammals their dung is much more conspicuous than the animal. An interesting example is the Amazonian manatee, which is extremely shy and difficult to spot at the water surface. However, it produces floating dung that is characteristic and easily spotted. Generally a simple count of the droppings in a known area will give an index of abundance over the indeterminate period over which the droppings have remained intact. Rainfall and other physical variables will greatly change this time period, as will biotic factors such as the abundance of dung beetles. There are two ways to overcome this problem, either remove all the droppings from the study area prior to starting the census (great work if you can get it) or mark some fresh dung and subsequently never count anything which looks older. Animal droppings are often highly clumped in their distribution and may conform to a negative binomial distribution. Statistical methods to compare densities are discussed by White and Eberhardt (1980).

The presence of mammals such as squirrels can be detected by the characteristic tooth marks they leave on the remains of their food such as nuts. A wide variety of feeding signs have been used in mammalian studies. An unusual example is the use by Turner (1975) of bite marks on cattle as an index of vampire bat, *Desmodus rotundus*, abundance.

Mammal density has also been monitored using hair samplers that snag hair from a passing animal. These can work in similar fashion to the way a barbed wire fence catches the wool of sheep or the hair may be retained on sticky tape placed inside a tube placed where the animals will use it as a runway (Suckling 1978). These devices are probably only useful for demonstrating the presence of a species at a locality for the quantity gathered will reflect many variables other than population size. However, DNA identification techniques may soon make it practicable to identify an individual from a single hair.

Footprints and runways may also be used. Mooty and Karns (1984) showed that the density of white-tailed deer inferred from tracks was correlated with the density inferred from pellet counts. Whatever type of sign is used, the data can be analyzed in two principal ways. The density can be expressed as counts per area of search or alternatively signs can be used instead of direct animal observations in distance sampling methods (see Chapter 5).

Chapter Eight

Estimating age and growth

Considerable research has been undertaken into estimating the age of plants and animals and this chapter can only introduce the subject. The age of an organism is an important variable. It is impossible to build life-tables or to analyze the population dynamics of a species without knowledge of the age structure of the population together with estimates of the age-specific mortality and fecundity rates.

Age determination can be difficult and some animal groups including leeches, flatworms, and sea anemones have never been satisfactorily aged. This chapter introduces the techniques that are widely used for major animal groups. While a method may be generally applicable to the members of a taxon the ease and accuracy with which it may be applied can vary greatly between species, so whenever possible the advice of someone with experience should be sought.

In the first part of this chapter the various methods for aging are introduced followed in the second part by a description of the commonly used growth models.

Using size–frequency histograms

It is common practice, particularly in animals that reproduce in synchrony once a year, to age group using a size–frequency histogram to identify the cohorts. Figure 8.1 shows as an example the length–frequency distribution of sprats captured in Bridgwater Bay in the Bristol Channel during the winter. The population comprises three age groups, which are clearly seen as three modes in the distribution. The animals in their first year of life (termed young of year or O-group in fish biology) have a modal length of about 50 mm, the fish in their second year of life, termed I-group, have a modal length of about 80 mm, and the fish in their third year of life, termed II-group, have a modal length of about 120 mm.

There are two major problems with this method: firstly, individual variation in growth can be considerable both within and between cohorts resulting in erroneous age assignment and, secondly, growth usually decelerates with age

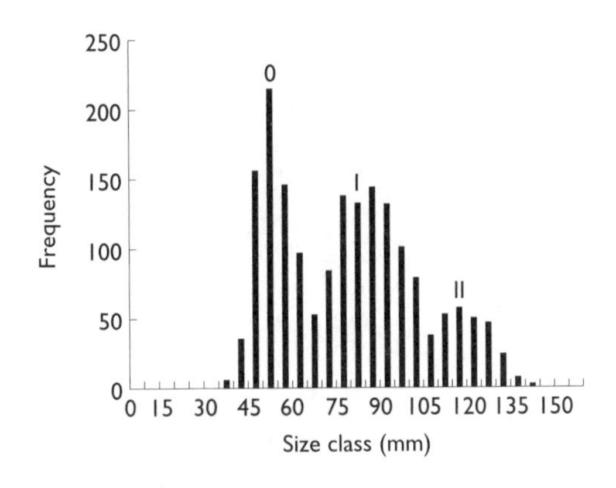

Figure 8.1 Size–frequency of sprat in the Bristol Channel, England showing three age groups.

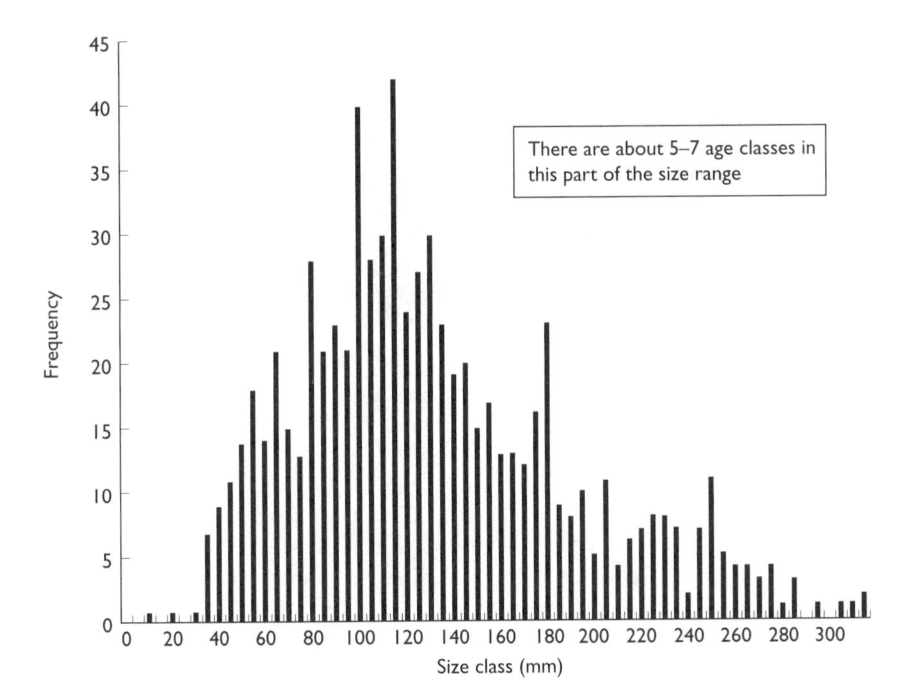

Figure 8.2 Size–frequency of flounder in the Bristol Channel, England. There are at least six age classes in this population, however the deceleration of growth with age combined with individual variation in growth and the lower abundance of older individuals makes the older age classes impossible to distinguish.

making it difficult or impossible to discriminate between the older age classes. Figure 8.2 shows the size distribution for flounder in the Bristol Channel. It is known from the examination of otoliths that this population comprises at least six age groups. However, the older age groups cannot be distinguished by size. An extreme example of these pitfalls is shown by the

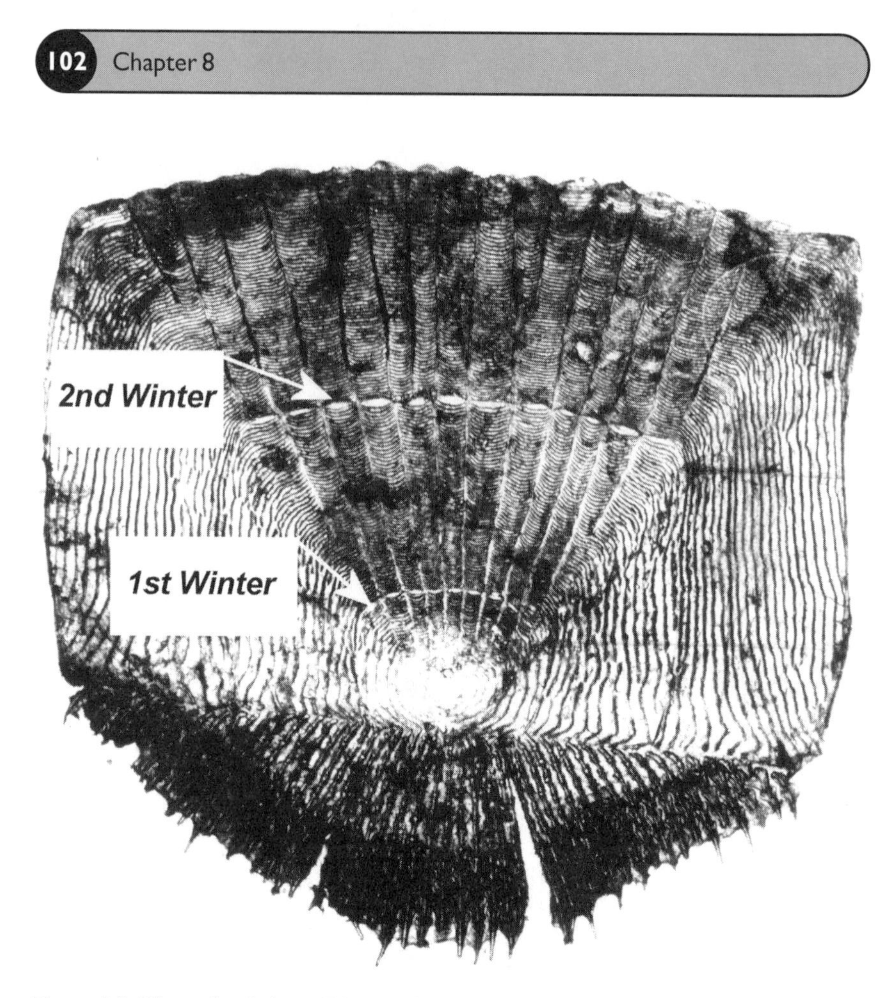

Figure 8.3 The scale of a bass, *Dicentrarchus labrax*, taken from a fish in the third winter of its life. The growth checks during the first and second winter are clearly seen as white lines.

study of Galinou Mitsoudi and Sinis (1995) on the long-lived bivalve *Lithophaga lithophaga* where it was found that individuals of 5.0 ± 0.2 cm in length ranged in age from 18 to 36 years. Size–frequency distributions can be used with confidence if they are supported by prior knowledge of the age–length relationship and individual variation (Goodyear 1997).

Using growth checks for age determination

In many organisms changes in the rate of growth leave marks on some part of the body that can be used to age them (Fig. 8.3). In temperate regions growth checks are usually associated with the winter. In the tropics they may be associated with the wet or dry season. The best-known example of such growth checks is tree rings, and their use to age and date wood has resulted in the development of the field of dendrochronology. In addition, tree rings are used to infer past changes in the climate. Growth rings can also be used to age and infer past growth in a wide number of animals. Some examples of their use are given below.

Annelids

In temperate regions annual growth rings are formed in the jaws of the polychaetes *Harmothoe imbricata*, *H. derjugini*, *Halosydna nebulosa*, *Hermilepidonotus robustus*, *Lepidonotus squamatus* and *Eunoe* sp. (Britaev & Belov 1993).

Insects

Cuticular bands, which are apparently daily growth layers, are present in some areas of the cuticle of many insects.

Bivalve molluscs

Bivalve molluscs gradually increase the thickness of the shell so that growth checks can show as a series of bands in suitably prepared cross-sections. In temperate regions the horse mussel, *Modiolus modiolus*, has alternating patterns of light (summer) and dark (winter) growth lines in the middle nacreous layer of the shell (Anwar *et al.* 1990). By counting such annual bands, Galinou Mitsoudi and Sinis (1995) were able to show that the piddock, *Lithophaga lithophaga*, could have a longevity of over 54 years.

Pulmonate molluscs

Growth bands may age these snails. Raboud (1986) aged *Arianta arbustorum* from the Swiss Alps using thin sections of the shell margins. In some habitats harsh winter conditions can give the shells of gastropods such as *Hydrobia ulva* clear winter growth checks. Terrestrial helicid snails are known to generally show bands caused by winter growth checks in the northern temperate regions and summer growth checks in Australia; these can be used for aging (Baker & Vogelzang 1988).

Fish

The allocation of fish to year class is normally accomplished by the examination of structures such as scales and otoliths. In temperate regions rapid summer and slow winter growth result in an annual pattern comprising a summer and a winter zone that differ in appearance (Fig. 8.3). The annulus is usually defined as the winter zone. A number of scales should be removed and examined from each fish. Age determinations may be made from direct observation, scale impressions, or photographs. A microfiche viewer is convenient for scale examination. Frequently encountered problems include difficulty in identifying the first annulus and indistinct patterns towards the edge of the scale. Otoliths are removed by dissection from the head of the fish and may be stored dry prior to examination. Examination for annuli may be undertaken using whole, baked, broken, or cross-sectioned otoliths depending on the species. Baking or burning enhances visibility of the annuli because the hyaline zones turn brown in contrast to the white opaque zones.

Reptiles and amphibians

Examination of skeletal structures are commonly used for reptile and amphibian aging. Francillon Viellot *et al.* (1990) aged the newts by counting annual growth rings in cross-sections of femurs and phalanges that had been stained with Ehrlich hematoxylin.

Mammals

The most widely applicable methods use the teeth and the most accurate estimates of the age at death are obtained by counting the incremental lines in the cementum of teeth.

Using biochemical methods for age determination

Crustaceans, particularly the long-lived crabs and lobsters, are difficult to age. However, Sheehy *et al.* (1994, 1996) have shown that the amount of lipofuscin in the left olfactory lobe cell mass of the brain of the Australian red-claw crayfish, *Cherax quadricarinatus*, or in the eyestalk ganglia of European lobster, *Homarus gammarus*, is positively correlated with age. The lipofuscin concentration was measured using confocal fluorescence microscopy and image analysis. Krill can also be aged by the amount of fluorescent pigment (Berman *et al.* 1989). In insects, Thomas and Chen (1989) reported that fluorescent pteridines in the compound eyes of adult screwworms, *Cochliomyia hominivorax*, increased in concentration through time and could be used for age determination. Similarly, Camin *et al.* (1991) reported that the homogenized head capsules of Mediterranean fruit flies, *Ceratitis capitat*, measured spectrofluorometrically showed significant differences in fluorescence from 0 to 28 days after emergence. This pteridine accumulation method has also been shown to work for the melon fly, *Bactrocera cucurbitae* (Mochizuki *et al.* 1993). Color change can be used for some insects, for example, the coloration of the bodies of adult dragonflies changes with age (Corbet 1962a,b).

Marking bony parts for growth studies

Various compounds can permanently stain bone, teeth, spines, scales, shell or otoliths as they are formed, creating a mark that can be used to measure the subsequent growth rate of the structure. These techniques are termed date banding or time stamps. Tetracycline antibiotics, because of their properties of localization in hard tissues such as bone, low toxicity, and fluorescence, have been used with success to mark many different animal groups. A bibliography of the marking of fish with tetracycline and its effects has been produced by Wastle *et al.* (1994). Oxytetracycline can also be administered in the food of mammals, e.g. red fox, *Vulpes vulpes*, striped shunks, *Mephitis mephitis*, and raccoon, *Procyon lotor*. Tetracycline staining can also be applied to the skeletal structures of lower animals such as the molluscs (Ekaratne & Crisp 1982), bryozoans (Barnes 1995), and sponges (Bavestrello *et al.* 1993).

Fish are probably the best organisms upon which to apply tetracycline staining to study growth. The basic protocol is to take a batch of young fish from a single cohort or age group and mark them by the addition of tetracycline to the water in their holding tank. Some of the individuals may then need to be sacrificed to obtain the length or weight of the cohort. The marked fish are then returned to the habitat. At some future time the habitat is resampled and the captured fish inspected for tetracycline marks, weighed, and measured again. The change in weight or length over a known time can be used to calculate the rate of growth and even fit growth models. Such marks can also be used to determine if fish or other animals are laying down a single growth check on their scales each year. In some habitats, particularly in the tropics or warmer temperate zones, there may be two or more growth checks associated with adverse conditions during periods of low water, high temperature, or low oxygen.

Describing growth mathematically

A great number of different mathematical expressions can be used to describe the growth of plants and animals. A small number of expressions that are reasonably simple and show many of the features observed have become widely used. These are briefly introduced below.

von Bertalanffy growth equation (Fig. 8.4a)

Let us assume that the rate of growth of an organism declines with size, then the rate of change in size may be described by:

$$\frac{dl}{dt} = K(L_\infty - l) \quad (8.1)$$

where K = the growth rate, l = the length of the organism, and L_∞ = the length which the organism would theoretically reach if it survived for ever.

Integrating this becomes:

$$l_t = L_\infty \left\lfloor 1 - e^{-K(t-t_0)} \right\rfloor \quad (8.2)$$

Von Bertalanffy derived this equation in 1938 from simple physiological arguments. The parameter t_0 is included to adjust the equation for the initial

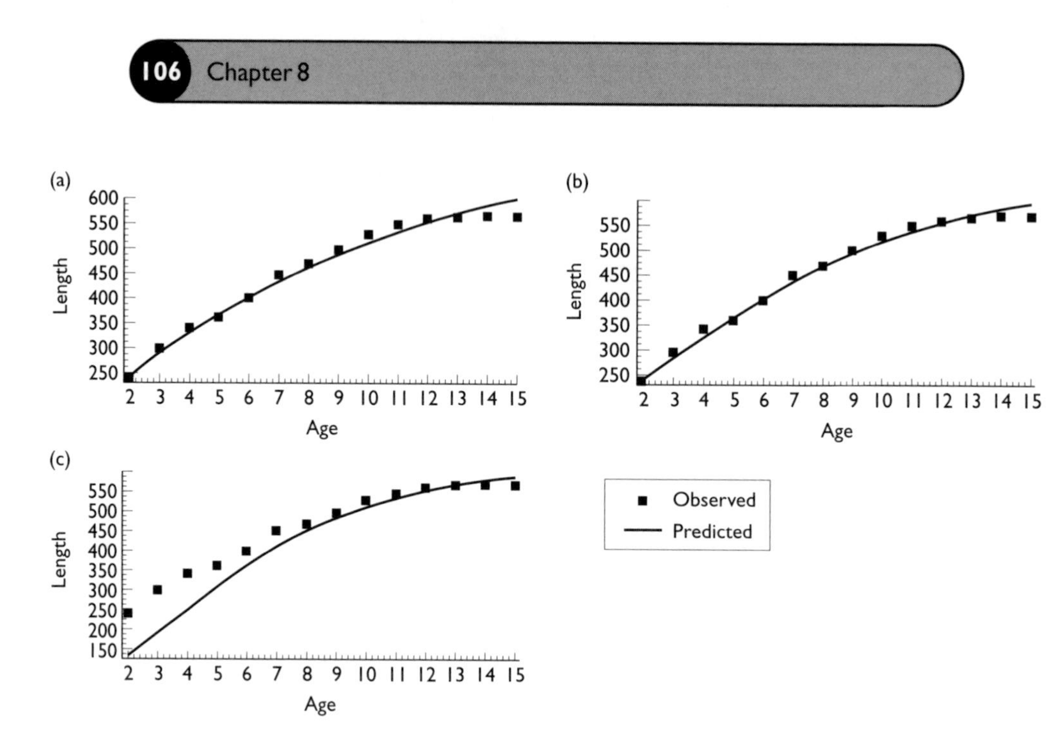

Figure 8.4 (a) Von Bertalanffy, (b) Gompertz, and (c) logistic growth curves fitted to length at age data for the bass, a common Mediterranean and Atlantic fish.

size of the organism and is defined as age at which the organisms would have had zero size. This equation is the most widely used growth curve and is especially important in fisheries studies.

Gompertz growth equation (Fig. 8.4b)

Let us assume that the rate of growth of an organism declines with size so that the rate of change in size may be described by:

$$\frac{\mathrm{d}\log l}{\mathrm{d}t} = K(\log L_{\infty} - \log l) \tag{8.3}$$

where l = length (size) and K = the growth rate.

After integration and some rearrangement:

$$l_t = L_{\infty}\mathrm{e}^{\mathrm{e}^{-K(t-I)}} \tag{8.4}$$

where L_{∞} = the length (size) at which growth is zero and I = the age at the inflection point for the curve.

Logistic growth equation (Fig. 8.4c)

Let us assume that the rate of growth of an organism initially accelerates then declines with size so that the rate of change in size may be described by:

$$\frac{dl}{dt} = l(K - \delta l) \tag{8.5}$$

where l = length (size), K = the growth rate, and δ is a term which expresses the rate at which growth declines with size.

After integration and some rearrangement:

$$l_t = \frac{L_\infty}{e^{-K(t-I)} + 1} \tag{8.6}$$

where L_∞ = the length (size) at which growth is zero and I = the age at the inflection point.

Fitting growth curves

Fitting any of the above equations to your data will require the use of a computer program. While specialist software to undertake this task can be used, it is also possible to use the regression packages in some standard statistical analysis software. However, in this case you may need to specify the equation as many packages do not include these equations as standard forms. As an example of the type of fits that can be obtained for the growth curves described above, length at age data for a fish called the bass was fitted to all three of the growth equations described above using a program called SIMPLY GROWTH (Pisces Conservation Ltd, www.irchouse.demon.co.uk). It can be seen in Figure 8.4 that for this species the von Bertalanffy and Gompertz curves gave a better fit than the logistic curve.

Chapter Nine

Life-tables and population budgets

Life-tables are a useful way of summarizing population data to aid our understanding of the population biology of a species and have long been used by actuaries for determining the expectation of life of an applicant for insurance. An example of an actual life-table is shown in Table 9.1. Using a life-table you can gain considerable insight into the dynamic properties of the population and the life-history strategies of the individuals. There are two types of life-table and the one that is constructed will depend on the properties of the organism under study and the type of data you can collect.

An age-specific (or horizontal) life-table is based on the fate of a real cohort, normally the members of a population belonging to a single generation. An age-specific life-table can therefore be constructed for short-lived organisms that breed in synchrony so that a well-defined cohort can be followed through time. You cannot create an age-specific life-table for a redwood tree although, given access to records, you might be able to create one for all of the peoples born in a selected village between say 1910 and 1920.

A time-specific (or vertical) life-table is based on the fate of an imaginary cohort found by determining the age structure, at one instant in time, of a sample of individuals from what is assumed to be a stationary population with considerable overlapping of generations, i.e. a multistage population. Such models are frequently constructed for fish populations. It might even be possible to construct one for the population of a particular species of tree in a forest, but it will only be possible if you can age the individuals as age determination is a prerequisite for the construction of time-specific life-tables (see Chapter 8).

The ecologist is generally interested in a life-table that records the actual number of individuals entering into each stage of the life cycle. Therefore, in much work on populations, the table lists the actual population number at different stages and records the action of the different mortality factors. Such tables are termed budgets.

The construction of a budget

In practice, it is only possible to construct a budget for a short-lived species that can be regularly sampled or observed. Appropriate organisms are the inhabitants of ephemeral pools particularly crustaceans such as fairy shrimps, *Tripos* sp., cladocerans, or ostracods. These may pass through the entire life cycle in 40 days or less and the different developmental stages differ in size and morphology. Because the onset of development is also synchronized by the rehydration of the pools, a series of regular samples will follow the development of a cohort. However, not all the resting eggs may hatch at the same time and this can complicate the calculations. Given more time other organisms such as insects and annual plants can also be studied.

To construct a budget you need to calculate the number of individuals passing through each of the different stages in the life history. The degree of synchronization in the life cycle is an important factor affecting the ease or difficulty of this step. The ideal situation is when there is a point in time when all the individuals of one generation are in a given stage; a census at this time will provide the number required for the life-table. Unfortunately, this is often not the case because of differences in the rate of development, time of birth, etc. When this occurs it is necessary to estimate the number passing through the stage by taking a succession of population estimates of the number in the stage and in some way estimating the total from the resulting time series.

Calculating the number entering a stage in the life history

The estimation of the number passing through a particular stage in the life history is frequently rather difficult. Manly (1990) tabulates 23 methods that use stage–frequency data for estimating some combination of the numbers entering a stage, stage duration, and stage-specific survival rates. These methods vary in their assumptions and requirements. Many of these are mathematically complex and require the use of a computer. Of the early and computationally less demanding methods, Southwood's graphical method is the crudest and simplest but it gives reasonable estimates of the numbers entering a stage if mortality, which may be heavy, occurs entirely at the end of the stage. This may be reasonable for many organisms as it is the changes linked with moulting and morphological change which may make additional demands upon the physiology and increase vulnerability to disease and predators. For example, even the toughest crab is highly vulnerable after moulting before the shell has hardened. If a constant mortality rate applies throughout the stage then the estimate will overestimate the numbers at the medium age of the stage. Successive estimates are plotted on graph paper, most conveniently allowing one square per individual and per day (Fig. 9.1). The points are joined up and the number of squares under the line counted; this total is then divided by the mean developmental time under field conditions to give the estimate of the numbers reaching the median age for the stage.

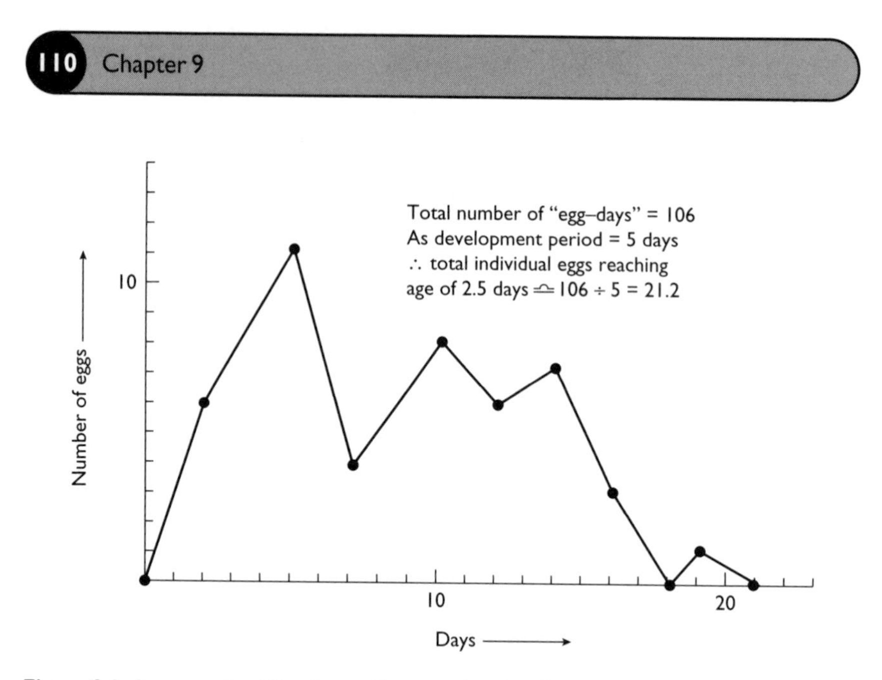

Figure 9.1 An example of Southwood's approximation for the estimation of the total number of individuals in a stage. The number of squares under the line is counted and divided by the estimated stage duration. (From Southwood & Henderson 2000.)

The description of budgets and life-tables

Having created a series of estimates of the number of individuals entering each stage of the life cycle it is then necessary to present the data in a manner that will allow the patterns in the data to be revealed.

Survivorship curves

The simplest description of a budget is the graphical representation of the fall-off of numbers with time – the survivorship curve. The numbers living at a given age (l_x) are plotted against the age (x); the shape of curve will describe the distribution of mortality with age. Slobodkin (1962) recognized four basic types of curve (Fig. 9.2): in type I mortality acts most heavily on the old individuals; in type II (a straight line when the l_x scale is arithmetic) a constant number die per unit of time; in type III (a straight line when the l_x scale is logarithmic) the mortality rate is constant; and in type IV mortality acts most heavily on the young stages.

Fish and many marine organisms with high fecundities show type IV curves. In insects mortality often occurs at particular stages so that survivorship curves show a number of distinct steps.

The life-table and life expectancy

An interesting variable that can be calculated once you have created a budget is the life expectancy at age. In human terms, this is the number of additional

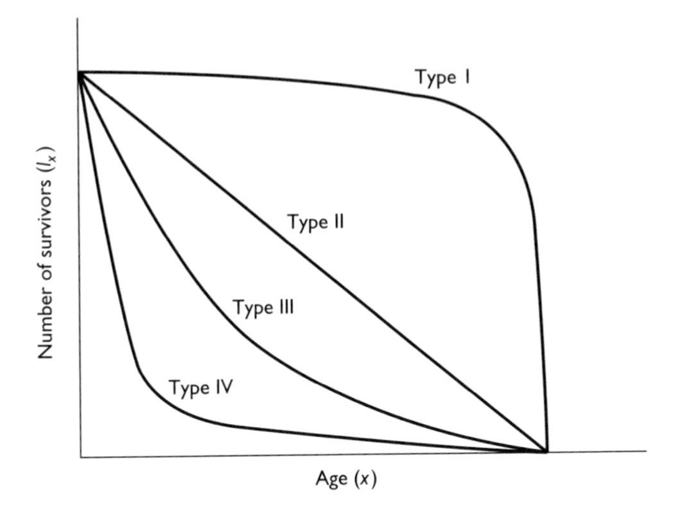

Figure 9.2 The four types of survivorship curve. (After Slobodkin 1962.)

Table 9.1 An example life-table. The barnacle, *Balanus glandula*, was studied by Connell (1970) for every year between 1959 and 1968 on the shore of San Juan Island, Washington.

Age (yr) x	Observed no. of barnacles alive each year	No. surviving at start of age interval x, l_x	No. dying within age interval x to $x+1$, d_x	Rate of mortality, q_x	Mean expectation of further life for animals alive at the start of age x
0	142	1000	563	0.563	1.58
1	62	437	198	0.453	1.96
2	34	239	98	0.410	2.17
3	20	141	32	0.227	2.34
4	15.5 (estimated)	109	32	0.294	1.88
5	11	77	31	0.403	1.45
6	6.5 (estimated)	46	32	0.696	1.11
7	2	14	0	0.000	1.50
8	2	14	14	1.000	0.50
9	0	0	—	—	—

years that a person of known age can on average expect to live. When building a life-table for this purpose it is convenient if the initial number at birth is set to a fixed number, often 1000. Table 9.1 shows an example of a life-table presented in this way for which life expectancy has been calculated.

Table 9.1 is constructed with the following columns:

x The pivotal age for the age class in units of time (days, weeks, years).

l_x The number surviving at the beginning of age class x (often out of 1000 originally born).

Table 9.2 A hypothetical life-table constructed to illustrate the calculations. (After Southwood & Henderson 2000.)

x	l_x	d_x	L_x	T_x	e_x	$1000q_x$
1	1000	300	850	2180	2.18	300
2	700	200	600	1330	1.90	286
3	500	200	400	730	1.46	400
4	300	200	200	330	1.10	667
5	100	50	75	130	1.30	500
6	50	30	35	55	1.10	600
7	20	10	15	20	1.00	500
8	10	10	5	5	0.50	1000

d_x The number dying during the age interval x.
e_x The expectation of life remaining for individuals of age x.

In practice the table may have two further columns (Table 9.2) to facilitate the calculation of the expectation of life as follows:

1 The number of animals alive between age x and $x+1$ is found. This is calculated using:

$$L_x = \int l_x d_x \approx \frac{l_x + l_{x+1}}{2} \tag{9.1}$$

2 The total number of animals x age units beyond the age x, which is given by:

$$T_x = L_x + L_{x-1} + L_{x-2} \ldots L_w, \tag{9.2}$$

where w = the last age. In practice it is found by summing the L_x column from the bottom upwards.

The expectation of life is theoretically:

$$e_x = \frac{\int_x^w l_x d_x}{l_x} \tag{9.3}$$

and is therefore given by:

$$e_x = \frac{T_x}{l_x} \tag{9.4}$$

When the survivorship curve is of type I (Fig. 9.2), e_x will decrease with age; it will be constant for type II and it will increase for types III and IV.

A further column is sometimes added to life-tables: the mortality rate per age interval (q_x) usually expressed as the rate per 1000 alive at the start of the interval:

$$1000q_x = 1000\frac{d_x}{l_x} \tag{9.5}$$

Life and fertility tables and the net reproductive rate

To study the rate of change in populations we need to study the birth rate (natality) in addition to the death rate (mortality). To construct such a table you will need to record the number of viable female offspring produced per female over different periods of her adult life. We are no longer concerned with males because they do not produce offspring and this immediately brings into focus one of the most interesting evolutionary questions, what are males for? Fertility tables are certainly easier to produce for parthenogenetic species such as many freshwater microcrustaceans.

A life and fertility table is constructed by preparing a life-table with x and l_x columns as before, except that the l_x column now refers only to females and their survival is expressed as the proportion of an initial population that is still alive (Table 9.3). An age-specific fertility (often termed fecundity) column (m_x) is added to the table and that records the number of living females born per female in each age interval. In many cases it is not possible to sex newborns or young, in which case a 1:1 sex ratio is often assumed.

Columns l_x and m_x are then multiplied together to give the total number of female births (female eggs laid) in each age interval (the V_x column).

The number of times a population will multiply per generation is described by the net reproductive rate R_0, which is:

$$R_0 = \int_0^\infty l_x m_x d_x = \sum l_x m_x \quad (9.6)$$

Thus from Table 9.3, $R_0 = 2.94$; this net reproductive rate may be expressed in another way as the ratio of individuals in a population at the start of one generation to the numbers at the beginning of the previous generation. Thus,

$$R_0 = \frac{N_{t+\tau}}{N_t} \quad (9.7)$$

where τ = generation time.

If R_0 is greater than 1 the population is increasing, if exactly 1 it is stable, and if less than 1 it is declining.

Table 9.3 Life and fertility table for the beetle, *Phyllopertha horticola*. (Modified from Laughlin 1965.)

x (in weeks)	l_x	m_x	$l_x m_x$ (V_x)
0	1.00	—	Immature
49	0.46	—	Immature
50	0.45	—	Immature
51	0.42	1.0	0.42
52	0.31	6.9	2.13
53	0.05	7.5	0.38
54	0.01	0.9	0.01

Calculating population growth rates

The growth rate of a population, r, is defined by the equation:

$$\frac{dN}{dt} = rN \tag{9.8}$$

where N is the number of individuals at any given time (t), which on integration gives:

$$N_t = N_0 e^{rt} \tag{9.9}$$

If the environment does not limit growth and the age structure is at equilibrium, then r will remain constant and if the value is positive the population will grow exponentially.

The maximum value of the parameter r that is possible for the species under the given physical and biotic environment is denoted as r_m and is variously termed the intrinsic rate of natural increase, the Malthusian parameter, or the innate capacity for increase, and is related to the finite capacity for increase λ, by the expression $e^r = \lambda$. The intrinsic rate of natural increase, r_m, is of value as a means of describing the growth potential of a population under given climatic and food conditions.

When a stable age distribution has been achieved, but the population is still growing in an unlimited environment, and given l_x, m_x values from a life-table, r_m may be calculated using the expression:

$$\sum e^{-r_m x} l_x m_x = 1 \tag{9.10}$$

by numerical means.

However, this is not straightforward without the aid of a computer so the following approximation for r_m has long been used by insect ecologists:

$$r_c = \ln R_0 / T_c \tag{9.11}$$

where T_c = cohort generation time, the mean age of the females in the cohort at the birth of female offspring or the pivotal age where $l_x m_x = 0.5\ R_0$.

Chapter Ten

Alpha diversity, species richness, and quality scores

In this chapter we will focus on the measurement of species diversity. In general terms this can be defined as a measure of the number of different species in an area of habitat. In practice, ecologists use the term species richness for an estimate of the total number of species present and the term diversity index for some measure of species number that also takes into account the relative abundance of the species present. You can never truly know the number of species living in a habitat, partly because of sampling limitations but also because it is not constant. The natural world is characterized by a constant flux of migrants and many species move between habitats seasonally or as they develop. It is also important to remember that species may remain hidden as seeds, spores or resting eggs waiting to develop when conditions change. The sudden growth and flowering of poppies following disturbance to the soil or a population explosion of daphnia or ostracods in temporary ponds after rain are good examples. These hidden species are clearly an important component of the diversity but will rarely be included in any field survey of diversity undertaken during typical conditions.

Given that our samples can only be a partial measure of the total number of species in a habitat it is still useful to obtain comparative measures for different localities. When comparing different habitats only selected taxonomic groups of plants or animals are studied, e.g. woody higher plants, spiders, or butterflies. When selecting groups for study you should ensure that you have a clear reason for your choice. The groups chosen for study may be selected for one or more of the following reasons:

1 They are ecologically dominant in the region under study and include species that are "keystone" species.
2 They are easy to catch, observe, and identify.
3 They include members that are particularly sensitive to changes in conditions or disturbance.
4 They form a group that is particularly characteristic of the habitat under study even if they are not particularly common.
5 They include rare or particularly valued species.
6 They are commercially valuable.

The data from which species richness or diversity can be calculated may be collected using almost any of the sampling approaches described in Chapters 3 and 6. While diversity indices require quantitative data on the relative abundance between species, some species richness estimators can be calculated using presence-absence data.

Different types of diversity

Diversity is one of those commonsense ideas that prove elusive and multi-faceted when precise quantification is sought. A useful classification, due to Whittaker (1972), is:

Alpha(α) diversity – the diversity of species within a community or habitat.

Beta(β) diversity – a measure of the rate and extent of change in species along a gradient, from one habitat to others.

Gamma(γ) diversity – the richness in species of a range of habitats in a geographic area (e.g. island); it is a consequence of the α diversity of the habitats together with the extent of the β diversity between them.

Thus α and γ diversity are qualities that simply have magnitude and could, theoretically, be described entirely by a single number (a scalar). By contrast, β diversity is analogous to a vector as it has magnitude and direction. Their descriptions therefore require different approaches. This chapter will consider α diversity and the following chapter β diversity.

Magurran (1988) gives a useful review of diversity and its meaning.

Estimating species richness

Species inventories for particular habitats or localities are frequently required for purposes such as conservation management. Because a complete census is rarely feasible the community must be sampled and two important questions need to be addressed. Firstly, have sufficient samples been taken to adequately characterize the community? Secondly, given that sampling can never give a complete list of the species present, how can we estimate the total species number, S_{max}, for the locality? Neither of these questions is easily answered, but below I introduce two simple approaches that are commonly used and offer insight into the types of approach that can be used.

Have sufficient samples been taken – the species accumulation curve

The plot of the cumulative number of species, $S(n)$, collected against a measure of the sampling effort (n) is termed the species accumulation curve. As the sampling effort increases the rate of discovery of new species will decline and so the accumulation curve should gradually stop rising. The change in the rate of acquisition of new species is a useful measure of the adequacy of your

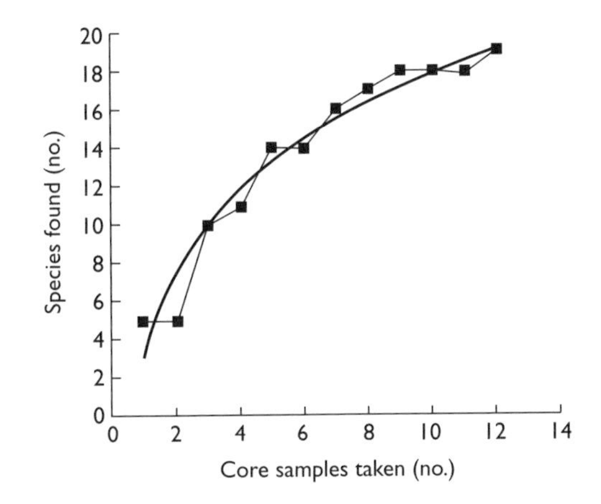

Figure 10.1 The species accumulation curve for the invertebrate fauna of a beach in Southampton Water, England, sampled by coring. (Graphical output produced by the program SPECIES RICHNESS AND DIVERSITY II (Pisces Conservation Ltd, www.irchouse.demon.co.uk).)

sampling. The sampling effort can be measured in many different ways; some examples are the number of quadrats or cores taken, total number of animals handled, hours of observation, or volume of water filtered. The species accumulation curve will only prove useful if applied to a defined habitat or area that is reasonably homogeneous.

Figure 10.1 shows the species accumulation curve for the number of species found in a series of 6.5-cm-diameter core samples of mud collected from a polluted beach in Southampton Harbour; the actual data are presented in Table 10.1. Straight lines connect the data points, but to illustrate the declining rate of accumulation of new species a smooth curve has been fitted to the data (how this was carried cut will be described later in this section). This graph immediately suggests that if more cores were to be taken further species would be found and the line is clearly still rising. However, the fact that the rate of increase is clearly declining would suggest that sufficient effort had been expended to give us confidence that the majority of species present have been recorded. If the species accumulation curve had shown an approximately constant increase in species number with effort we would conclude that more samples should have been taken.

Figure 10.2 illustrates the shape of a species accumulation curve when high sampling effort has been expended. It shows the cumulative species number of fish recorded from monthly samples collected between 1981 and 1995 at Hinkley Point in the lower Severn Estuary and the cumulative number of species of Heteroptera against the number of individuals collected over several years from oak trees. The sampling method used for fish, described by Henderson and Holmes (1991), was to use the power station cooling water pumps as a sampler and thus each month the volume of water filtered was a

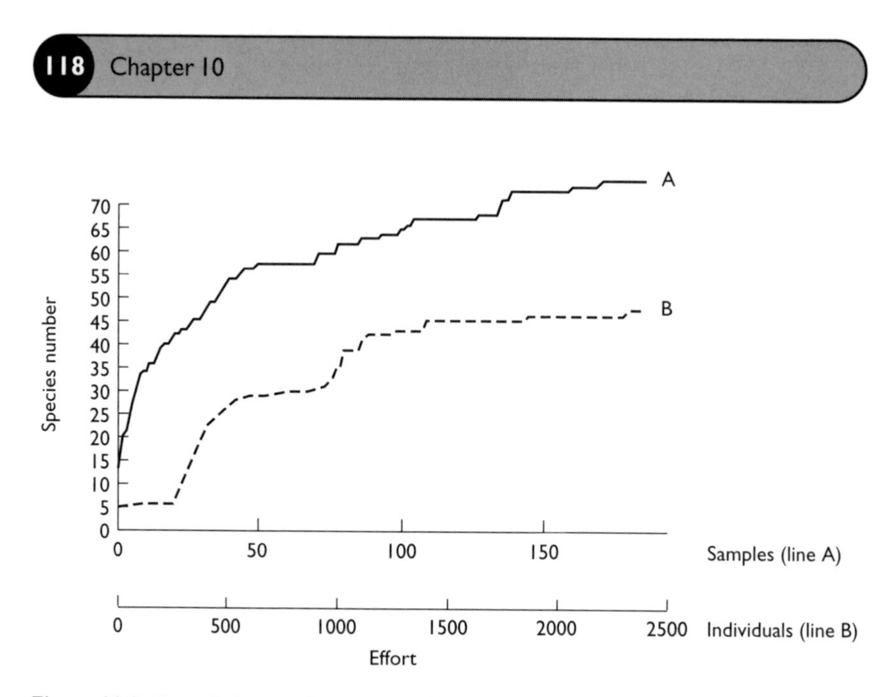

Figure 10.2 Cumulative species number of: line A, fish recorded from monthly samples collected between 1981 and 1995 at Hinkley Point in the lower Severn Estuary (after Henderson (1989), with more recent data from Pisces Conservation Ltd); line B, Heteroptera from oak trees (after Southwood 1996).

constant $3.24 \times 10^5\,m^3$, giving a total volume sampled to the end of 1995 of about $5.18 \times 10^7\,m^3$. This is a truly gigantic sampling volume when you realize that 1 cubic meter of seawater weighs just over 1 metric ton. As is normal when sufficient sampling effort is expended, the curve is asymptotic (gradually levels off). Yet, even given this exceptionally high sampling effort, new species are still occasionally caught. The inshore fish fauna of the North Eastern Atlantic is well known and Henderson (1988) argues that the total species number at this latitude in fully marine waters is about 80, although offshore or deep-water species may occasionally appear. Likewise with the oak Heteroptera, when a limited number of trees in two nearby woods were sampled by chemical knockdown, which is the most effective sampling technique for arboreal insects, yet more new species were found.

The order in which samples were taken alters the shape of a species accumulation curve both because of random error and sample heterogeneity. In Figure 10.1 for example the first two samples were from a region of the beach that was particularly species poor. In the case of the Hinkley data (Fig. 10.2), which is a time series, more fish species were caught during the autumn than at other times of the year so an accumulation curve commencing in September will rise faster than one starting in January. Further examination of Figure 10.2 shows no increase in species number between samples 44 and 76. These samples were collected during an exceptionally cold period in the mid 1980s when the inshore fish community was much reduced. To eliminate features caused by random or periodic temporal variation the sample order

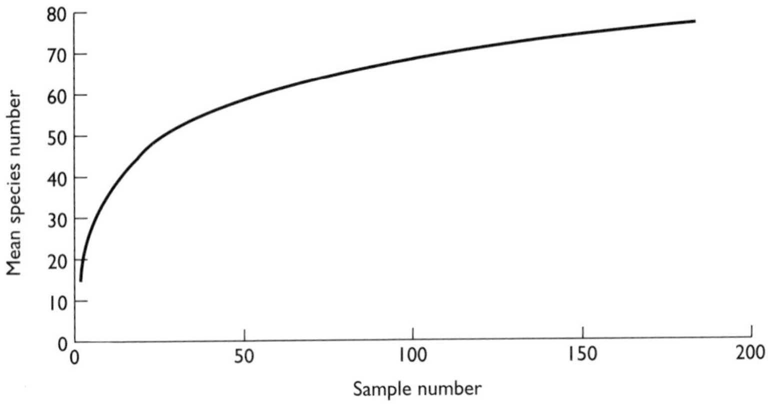

Figure 10.3 The average species accumulation curve from 20 randomizations of the Hinkley Point species accumulation curve shown in Figure 10.2. The calculations (and graphical output) were performed using the program SPECIES DIVERSITY AND RICHNESS II (Pisces Conservation Ltd, www.irchouse.demon.co.uk).

can be randomized. A useful procedure is to randomize the sample order r times and calculate the mean and standard deviation of $S(n)$ over the r runs (Colwell & Coddington 1994). This was how the smooth curve in Figure 10.1 was calculated. This approach can also be applied to samples collected along transects provided they are from a single community. As an example of how this averaging procedure leads to the production of an essentially asymptotic accumulation curve given a large number of samples, the average of 100 randomizations of the Hinkley Point data is shown in Figure 10.3. After about 50 samples had been taken about 60/80 (75%) of the total fish species complement had been recorded. For many studies this level of coverage is probably quite acceptable. Note here that the species number never actually becomes constant. This is because of the slow, but almost constant rate of arrival of rare migrants from other marine habitats.

Using the species accumulation curve to estimate total species richness

If a suitable model can be found to describe the species accumulation curve then it would be possible to estimate the total species complement of the habitat. The asymptotic behavior of the accumulation curve can be modeled as the hyperbola:

$$S(n) = \frac{S_{max} n}{B + n} \tag{10.1}$$

where S_{max} and B are fitted constants. This is the Michaelis–Menten equation used in enzyme kinetics and thus there is an extensive literature discussing the estimation of its parameters, which unfortunately presents considerable statistical difficulties. This approach is only likely to give sensible answers if the accumulation curve is clearly decelerating and the likely value of the asymptote can be estimated by eye. For this reason, the approach discussed below is more likely to be useful.

Nonparametric estimators of total species richness

This will give both a measure of the completeness of the inventory and also allow comparison with the species richness of other localities. Many different approaches have been suggested to estimate S_{max} at present. No clear consensus as to the best approach is available, although recent studies on parasite (Walther & Morand 1998) and fish communities (P.A. Henderson & A.E. Magurran, personal communication) suggest that a method developed by Chao performs well in terms of the reliability of the estimate given the effort required to obtain the data. As this offers a particularly simple approach to the calculation of S_{max} it is the method introduced here. A variety of other methods are described in Southwood and Henderson (2000).

Given a sample in which the number of individuals belonging to each species has been counted, the Chao index can be calculated using the following formula:

$$\hat{S}_{max} = S_{obs} + (a^2/2b) \qquad (10.2)$$

where a and b are the number of species represented by one and two individuals respectively and S_{obs} is the actual number of species observed. The calculations are demonstrated in Table 10.1 using data collected from a polluted beach in Southampton Harbour. The data are the number of individuals in 6.5 cm diameter core samples of mud retained by a 0.5 mm sieve. In Table 10.1 the calculations have been performed for each core sample, however these values are estimates of the total number of species in each core whereas we are more often interested in the total number of species in the beach. This is estimated in the right-hand column using the summed totals from all the samples. This gives an estimated total species richness of 22.25, which can be rounded up to 23 species. An examination of Figure 10.1 suggests that this value is quite a reasonable estimate of the asymptote of the species accumulation curve, giving increased confidence that it is a fair estimate given the modest number of samples.

If you only have presence-absence data from a number of samples then you can still use the Chao estimator by defining a as the number of species only found in one sample and b as the number of species only found in two samples. As before S_{obs} is the actual number of species observed. Using the core data again, $S_{obs} = 20$, $a = 4$, and $b = 1$, giving an estimate of total species richness of $20 + 16/2 = 28$. Although higher than that obtained above using quantitative data it is still perfectly reasonable and can be calculated from data that may be far easier to obtain.

Table 10.1 Invertebrates obtained from core samples on a beach in Southampton Water used to illustrate the calculation of total species richness estimators.

	Core sample												Total
Species													
Cerastoderma edule					2	1	1	1					5
Abra nitida	1												1
Abra sp.			1			3	1						5
Littorina littorea					1								1
Hydrobia ulvae	7	16	7	29	5	11	1	2	0	0	5	6	89
Cyathura carinata	2	4	26	14	1	1	0	0	0	0	0	0	48
Microdentopterus sp.			1		1								2
Amphipoda indet.			2	1									3
Amphilochus neapolitanus			2	1	1		1			2			7
Corophium volutator												3	3
Corophium sp.							1						1
Jaera albifrons grp									1				1
Tubificoides	77	56	30	25	102	185	56	67	0	0	46	19	663
Hediste diversicolor	2	3				2					6	21	34
Nephtys hombergi								1					1
Sabellid indet.			10			1							11
Monayunkia aestuarina				1									1
Amphorete acutifrons					2								2
Cirriformia tenticulata							1	3	2	1	3	10	20
Anemone indet.			1		1	3							5
No. observed S_{obs}	5	4	9	6	9	8	7	5	2	2	4	5	20
No. of singletons, a	1	0	3	3	5	3	6	2	1	1	0	0	3
No. of doubletons, b	1	0	2	0	2	1	0	1	1	1	0	0	2
$a^2/2b$	0.5	0	2.25	—	6.25	4.5	—	2	0.5	0.5	0	0	2.25
Estimated total species S_{max}	5.5	4	11.25	—	15.25	12.5	—	7	2.5	2.5	4	5	22.25

Comparing species evenness or equitability

Equitability or evenness is the pattern of distribution of the individuals between the species. A sample of 100 individuals representing 10 species could consist of 10 individuals of each – as each species is equally represented this would be the most extreme level of equitability possible. At the other extreme, the sample might comprise 91 individuals of one species (the dominant) and one each of the other nine. This sample would be said to be extremely uneven or with low equitability. Equitability is an important part of the description of a community and has important applications in ecological monitoring because highly stressed environments often show low levels of equitability as the system becomes dominated by disturbance or pollution-tolerant species.

The best way to display and compare the equitability of samples is to plot log number of individuals (or abundance) of each species against the rank. The rank is simply a number that gives the position of the species in a table of abundance, so the most abundant has a rank of 1, the second most abundant

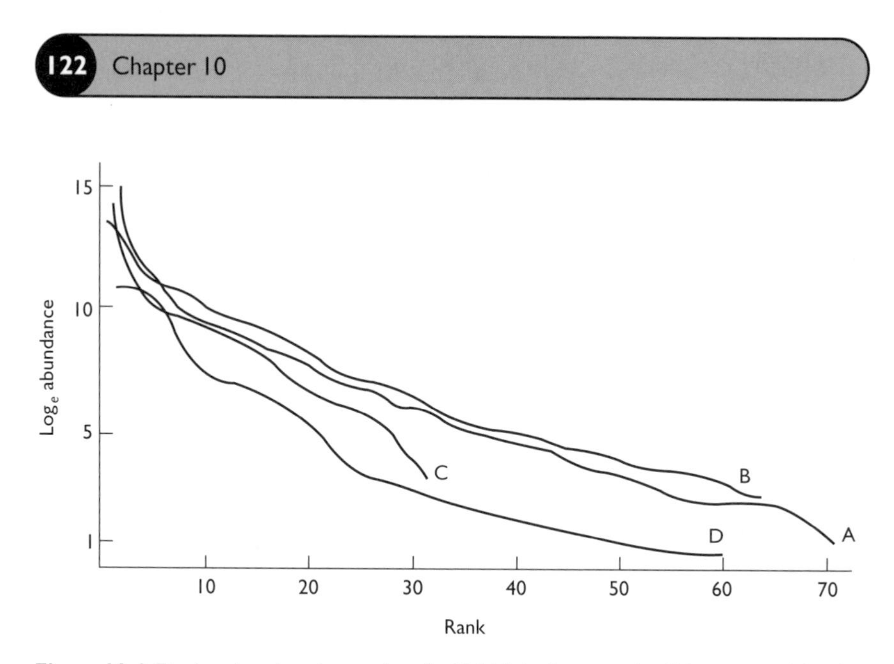

Figure 10.4 Rank order abundance plots for British inshore marine fish communities: A, Sizewell; B, Dungeness; C, Kingsnorth; D, West Thurrock. (After Henderson 1989.)

a rank of 2, and so on. An example showing the rank order–log abundance plots for British marine fish at different localities is shown in Figure 10.4.

The rank order–log abundance will often feature one or two highly abundant species occupying ranks 1 and 2 followed by the main series of species that approximately sit on a straight line of declining abundance. In large samples from high diversity habitats, there also tends to be a tail of species represented by between one and five individuals only. All of these features are represented to some extent in Figure 10.4, which shows that the communities at Sizewell and Dungeness were more equitable than those at Kingsnorth and West Thurrock. This is unsurprising as the latter two localities lie within the Thames Basin, a region greatly damaged by man.

The species abundance relationship can be fitted, not always very successfully, to a variety of models of which the best known are the geometric and logarithmic series, the log normal and the broken stick. Information on the fitting of these models can be obtained from Southwood and Henderson (2000).

Alpha diversity indices

Both the species richness and equitability of a data set may be summarized with a single number – a diversity index. It must be remembered that such a summary may not always have desirable properties. For example, the index will tend to be biased in favor of species number or equitability and thus may simply reflect a change in one of these quantities far more than the other. They can be useful in comparing localities or the same locality through time but are

Table 10.2 An example calculation of the Shannon–Wiener diversity index. The natural logarithm (log to base e) has been used; the index could also be calculated using standard base 10 logs if desired.

Species	No.	Proportion, p	$\log_e p$	$p \times \log_e p$
Abra nitida	1	1/89 = 0.0112	−4.488	−0.0503
Hydrobia ulvae	7	7/89 = 0.0787	−2.55543	−0.200
Cyathura carinata	2	2/89 = 0.0225	−3.795	−0.0854
Tubificoides	77	77/89 = 0.865	−0.145	−0.125
Hediste diversicolor	2	2/89 = 0.0225	−3.795	−0.0854
Total	**89**			−0.546

of no use for the comparison of different studies as they inevitably reflect the effort and techniques used by the workers. A process known as rarefaction, in which the size of the larger sample is reduced by a randomization process to that of the smaller, can produce samples of equivalent size from which diversity indices can be computed to compare the diversity of different sized samples. These techniques will not be discussed further here as it is assumed that all the samples are collected in a similar way.

There are numerous diversity indices reported in the literature and only the most frequently used are described below.

Shannon–Wiener* function (H)

This function was devised to determine the amount of information in a code, and is defined as:

$$H = -\sum_{i=1}^{S_{obs}} p_i \log_e p \qquad (10.3)$$

where p_i = the proportion of individuals in the ith species; H tends to increase with the number of species in the sample so it often gives little more insight than the species number. If you are using this index to compare the diversity of a series of samples also plot H against species number for each sample and see if the two variables are closely correlated.

The stages in the calculation of the Shannon–Wiener index for a small sample are laid out in Table 10.2. The final answer is $H = 0.546$ as there is a negative sign in front of the summation.

Simpson–Yule index (D)

This diversity index was proposed by Simpson (1949) to describe the probability that a second individual drawn from a population would be of the same species as the first. The statistic, C (or Y) is given by:

* Also referred to as the Shannon–Weaver function.

$$C = \sum_{i}^{S_{obs}} p_i^2 \qquad (10.4)$$

where, strictly,

$$p_i^2 = \frac{N_i(N_i - 1)}{N_T(N_T - 1)}$$

but is usually approximated as:

$$p_i^2 = \left(\frac{N_i}{N_T}\right)^2$$

where N_i is the number of individuals in the *i*th species and N_T the total individuals in the sample. The index is:

$$D = \frac{1}{C} \qquad (10.5)$$

and the larger its value the greater the equitability (range 1 to S_{obs}).

Berger–Parker dominance index

This index is simple, both mathematically and conceptually, being the ratio of the number of individuals in the sample belonging to the most abundant species, N_{max}, divided by the total number of individuals caught, N_T:

$$d = \frac{N_{max}}{N_T} \qquad (10.6)$$

So for the sample data given in Table 10.2, $N_{max} = 77$, $N_T = 89$, and $d = 77/89 = 0.865$. The index is not greatly influenced by the observed species number and is one of the best to use even if it does seem to lack mathematical sophistication!

McIntosh diversity measure

McIntosh (1967) suggested the dominance index:

$$D = \frac{N - U}{N - \sqrt{N}} \qquad (10.7)$$

where N is the total number of individuals in the sample. U is calculated as:

$$U = \sqrt{\sum n_i^2} \qquad (10.8)$$

and n_i is the number of individuals belonging to the *i*th species.

Using the sample data in Table 10.2, $N = 89$, $\sqrt{N} = 9.43$, and $U = \sqrt{5987}$ $(1^2 + 7^2 + 2^2 + 77^2 + 2^2 = 1 + 49 + 4 + 5929 + 4 = 5987) = 77.38$. Therefore, $D = 89 - 77.38 / 89 - 9.43 = 11.62/79.57 = 0.146$.

Comparing communities – diversity ordering

Different diversity indices may differ in the ranking they give to communities (Hurlbert 1971; Tóthméresz 1995). An example from Tóthméresz (1995) illustrates the point. Consider three imaginary communities with the following sets of species abundances for each of which diversity has been calculated using both Shannon–Wiener (H) and Simpson's (D) indices:
Community A: {33,29,28,5,5}, $H = 1.3808$, $D = 3.716$.
Community B: {42,30,10,8,5,5}, $H = 1.4574$, $D = 3.564$.
Community C: {32,21,16,12,9,6,4}, $H = 1.754$, $D = 5.22$.

Because $H(\mathrm{B}) > H(\mathrm{A})$ it could be argued B is the most diverse; however, as $D(\mathrm{A}) > D(\mathrm{B})$ the opposite conclusion could also be entertained. Communities such as A and B that cannot be ordered are termed noncomparable. Such inconsistencies are an inevitable result of summarizing both relative abundance and species number using a single number (Patil & Taillie 1979). Diversity profiles offer a solution to this problem by identifying those communities that are consistent in their relative diversity. To identify these communities it is necessary to produce an expression that can generate the various indices by changing the value of preferably a single parameter.

Perhaps the most generally useful expression of this type is that due to Renyi (1961), which is based on the concept of entropy and defined as:

$$H_\beta = \frac{\left(\log \sum_{i=1}^{s} p_i^\beta\right)}{(1-\beta)} \tag{10.9}$$

where β is the order ($\beta \geq 0$, $\beta \neq 0$), p_i the proportional abundance of the ith species, and log the logarithm to a base of choice, often e.

Hill (1973) used an almost identical index N_a which is related to H_β by the equality:

$$H_\beta = \log(N_a) \tag{10.10}$$

He demonstrated that N_a for $a = 0, 1, 2$ gives the total species number, Shannon–Wiener H and Simpson's D respectively. Thus by using different values of β, or "a", a range of diversity measures can be generated from the Renyi or Hill equations. To test for noncomparability of communities H_β is calculated for a range of β values and the results presented graphically. Figure 10.5 shows the diversity ordering of the three artificial communities, A, B, and C. This figure shows that C is always greater than A or B and thus can be considered to be more diverse. As A and B cross over one cannot be considered more diverse than the other by all reasonable measures.

We can summarize the application of the ideas and techniques introduced in this chapter by a series of numbered steps:

1 Plot graph(s) of log abundance on rank. It has always been important to examine the form of your data and this is easily done on a computer. Does the data form a straight line? Which are the species that depart most from it? Is there anything unusual in their biology (e.g. they could be vagrants)? These

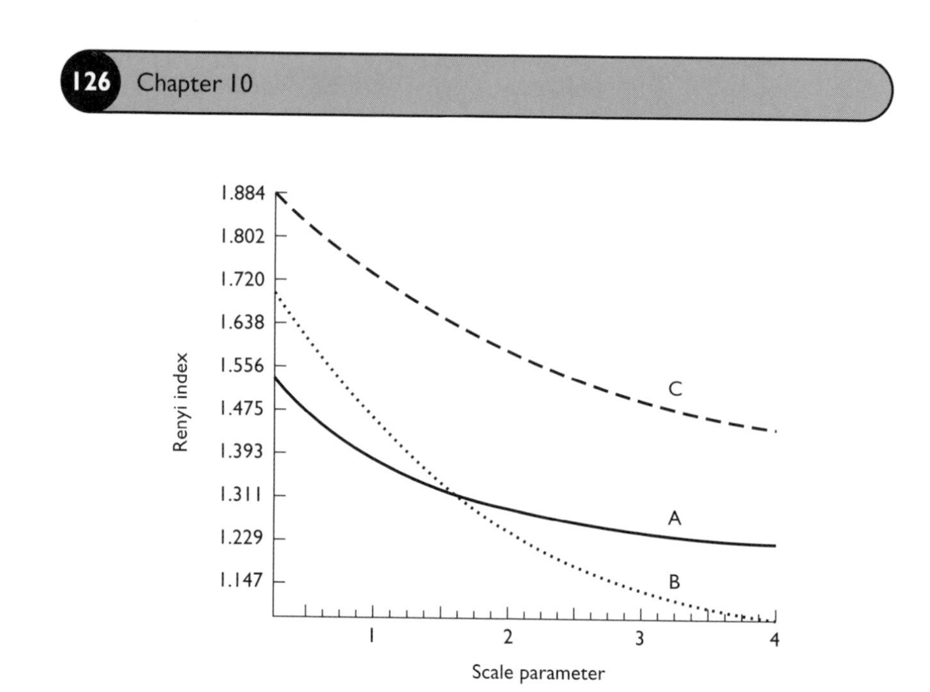

Figure 10.5 The diversity ordering of the three artificial communities, A, B, and C. Community C is the more diverse using any measure of diversity. It is not possible to state that community A is more or less diverse than community B because the relative magnitude of their diversities depends on the index used. (Graphical output produced by the program SPECIES RICHNESS AND DIVERSITY II (Pisces Conservation Ltd, www.irchouse.demon.co.uk).)

graphs may be an excellent manner of presenting the data for publication (e.g. Southwood *et al.* 1979; Henderson 1988).

2 Plot the species accumulation curves and calculate total species richness, S_{max}. The species accumulation curve gives insight into the sufficiency of the sampling effort. There is no best method to calculate S_{max} but the Chao method is as good as many and simple.

3 Calculate the Berger–Parker dominance index.

4 If sites are to be compared, undertake diversity ordering to ensure that they can be compared using a diversity index.

5 Calculate other indices if desired.

Habitat quality scores

Regular monitoring is frequently carried out to ensure that man has not adversely affected a habitat. It would often be too costly to undertake a full quantitative analysis of the flora or fauna present, so some system of scoring the major taxonomic groups present is devised which will give a measure of the general quality of the community and will indicate if any degradation has occurred. One of the best examples of such a quality scoring system is the Biological Monitoring Working Party (BMWP) Score for stream fauna used in the UK. Similar scoring schemes can be used in almost all parts of the world. The BMWP Score system is a method of assessing water quality using the

Table 10.3 BMWP scores for freshwater invertebrate families found in streams in northern Europe. The total score for a sample is obtained by summing the individual scores for the taxa present.

Siphlonuridae	10	Corduliidae	8	Dytiscidae	5
Heptageniidae	10	Libellulidae	8	Gyrinidae	5
Leptophlebiidae	10	Psychomyiidae	8	Hydrophilidae	5
Ephemerellidae	10	Philopotamidae	8	Clambudae	5
Potamanthidae	10	Caenidae	7	Helodidae	5
Ephemeridae	10	Nemouridae	7	Dryopidea	5
Taeniopterygidae	10	Rhyacophilidae	7	Elminthidae	5
Leuctridae	10	Polycentropodidae	7	Chrysomelidae	5
Capniidae	10	Limnephilidae	7	Curculionidae	5
Perlodidae	10	Neritidae	6	Hydropsychidae	5
Perlidae	10	Viviparidae	6	Tipulidae	5
Chloroperlidae	10	Ancylidae	6	Simuliidae	5
Aphelocheiridae	10	Hydroptilidae	6	Planariidae	5
Phryganeidae	10	Unionidae	6	Dendrocoelidae	5
Molannidae	10	Corophiidae	6	Baetidae	4
Beraeidae	10	Gammaridae	6	Sialidae	4
Odontoceridae	10	Platycnemididae	6	Piscicolidae	4
Leptoceridae	10	Coenagriidae	6	Valvatidae	3
Goeridae	10	Mesoveliidae	5	Hydrobiidae	3
Lepidostomatidae	10	Hydrometridae	5	Lymnaeidae	3
Brachycentridae	10	Gerridae	5	Physidae	3
Sericostomatidae	10	Nepidae	5	Planorbidae	3
Astacidae	8	Naucoridae	5	Sphaeriidae	3
Lestidae	8	Notonectidae	5	Glossiphoniidae	3
Agriidae	8	Pleidae	5	Hirudidae	3
Gomphidae	8	Corixidae	5	Erpobdellidae	3
Cordulegasteridae	8	Haliplidae	5	Asellidae	3
Aeshnidae	8	Hygrobiidae	5	Chironomidae	2
				Oligochaeta	1

families of insects and other aquatic invertebrates present. Each family present in a sample is given a score (Table 10.3) and the total score for samples of the fauna in a stream gives a measure of habitat quality. The family scores have been allocated by a group of experts so that the groups most sensitive to environmental disturbance are given the highest scores. Other variants on this theme include the Average Score Per Taxon (ASPT), which is simply the BMWP score divided by the number of scoring taxa in the sample.

Another approach, developed and widely used in North America, is the Index of Biological Integrity (IBI). This system of scoring habitats was originally devised by Karr (1991) to assess the condition of streams but has subsequently been adopted for other habitats such as uplands (e.g. Fore *et al.* 1996; Karr & Chu 1997). An IBI measures the degree to which the habitat is maintaining "a balanced, integrated, adaptive community of organisms having a species composition, diversity, and functional organization comparable to

Table 10.4 The criteria used by Karr (1991) in his original Index of Biological Integrity for small warm water streams in central Illinois and Indiana. Similar criteria, adapted to reflect the local fauna, are now widely used in North America.

Category	Metric
Species richness and composition	**1** Total number of fish species
	2 Number and identity of darter species
	3 Number and identity of sunfish species
	4 Number and identity of sucker species
	5 Number and identity of intolerant species
	6 Proportion of individuals as green sunfish (tolerant species)
Trophic composition	**7** Proportion of individuals as omnivores
	8 Proportion of individuals as insectivorous cyprinids (minnows)
	9 Proportion of individuals as top carnivore
Fish abundance and condition	**10** Number of individuals in sample
	11 Proportion of individuals as hybrids
	12 Proportion of individuals with disease, tumors, fin damage, or skeletal anomalies

that of natural habitat of the region" (Karr & Dudley 1981). For terrestrial systems birds have been used instead of fish as the focal taxonomic group.

The original version used 12 biological measures (termed metrics) that reflected fish species richness and composition, trophic structure, and fish abundance and condition (Table 10.4). To calculate the IBI for a site each metric is given a score and the sum of the scores for the 12 metrics is the IBI. A metric scores 5 points if the stream has characteristics expected for a fish community with little human influence, it scores 1 if it is appreciably different from the expected natural condition, and it scores 3 if the characteristics are intermediate. Thus the IBI ranges from 60 for pristine habitats to 12 for severely impacted streams.

One reason why IBIs have not been more widely applied in some areas in western Europe is probably because the extent of human interference is so great it is impossible to define the reference habitat. However, the use of more than just diversity when determining the quality of a habitat is clearly desirable.

When undertaking a survey, it may be appropriate to devise your own index of community quality. It is also an interesting research project to design an appropriate index. When creating an index you should consider carefully the sensitivity of the index to detect change and the ease with which it can be used.

Chapter Eleven

Species along environmental gradients – beta diversity

Beta (β) diversity measures the turnover in species composition along transects and is particularly applicable to the study of environmental gradients. To calculate β diversity you will need to collect information on the presence of species at a number of localities along a gradient. The information may comprise either presence-absence or quantitative abundance data. β diversity measures two attributes, the number of distinct habitats within a region and the replacement of one species by another between disjoint parts of the same habitat. Used together, alpha (α) and β diversity can give an assessment of the diversity of an area (Routledge 1977; Southwood *et al.* 1979). Many of the methods were originally developed and extensively used in work on plant and bird ecology. As with α diversity, a variety of indices have been proposed and no single measure is clearly superior for all types of study.

The following six indices for presence-absence data were considered by Wilson and Schmida (1984).

1 Whittaker's β_w:

$$\beta_w = \frac{S}{\alpha - 1} \tag{11.1}$$

where S = the total number of species and α = the average species richness of the samples (not the alpha diversity!). All samples must have the same size (or sampling effort).

2 Cody's β_c:

$$\beta_c = \frac{g(H) + l(H)}{2} \tag{11.2}$$

where $g(H)$ = the number of species gained and $l(H)$ = the number lost moving along the transect.

3–5 Routledge's β_R, β_I, and β_E

$$\beta_R = \frac{S^2}{2r + S} - 1 \tag{11.3}$$

where S = the total species number for the transect and r = the number of species pairs with overlapping distributions.

Assuming equal sample sizes:

$$\beta_I = \log(T) - \left[\left(\frac{1}{T}\right)\sum e_i \log(e_i)\right] - \left[\left(\frac{1}{T}\right)\sum \alpha_i \log(\alpha_i)\right] \qquad (11.4)$$

where e_i = the number of samples along the transect in which species i is present, α_i = the species richness of sample i, and $T = \Sigma e_i$.

The third of Routledge's indices is simply

$$\beta_E = \exp(\beta_I) - 1 \qquad (11.5)$$

6 Wilson and Schmida's β_T:

$$\beta_T = \frac{[g(H) + l(H)]}{2\alpha} \qquad (11.6)$$

where the parameters are defined as for β_c and β_w.

Based on an assessment of the essential properties of a useful index (ability to detect change, additivity, independence of the size of α, and sample size), Wilson and Schmida (1984) concluded that β_w was best. Harrison *et al.* (1992) apply this measure, in modified form, to the study of latitudinal gradients in a variety of aquatic and terrestrial groups.

Example calculations

Part of the Garraf vegetation data set of Kent and Coker (1992) is used to illustrate the calculation of β diversity measures. The samples were collected along a transect in northeast Spain, near Barcelona using a 5 m × 5 m quadrat, a relatively large area suitable for the shrubby structure of the vegetation. Only sites with a northern aspect are included here. The data were originally semi-quantitative with the percentage cover of each species recorded; in Table 11.1 only presence-absence is recorded as this is all that is required for the calculation of β diversity.

Whittaker's β_w

The total number of species (S) = 15, the average species richness $(\alpha) = (12 + 8 + 11 + 10 + 6 + 9 + 8 + 10)/8 = 9.25$. Thus $\beta_w = 15/(9.25 - 1) = 1.82$.

Cody's β_c

We compare the beginning and end of the transect, quadrats q1 and q8. There are 12 species in q1 and five of these are not present in q8. Therefore the number of species lost, $l(H) = 5$. There are three species not present in q1 (*Cortaderia selloana*, *Sedum* sp., *Smilax asper*) that are present in q8. Therefore the number of species gained, $g(H) = 3$.

Putting these values into the equation, $\beta_c = (5 + 3)/2 = 4$.

Table 11.1 The Garigue vegetation data collected along a transect near Garraf, northeast Spain.

Species	q1	q2	q3	q4	q5	q6	q7	q8	Occurrence (e)	ln(e)	e × ln(e)
Brachypodium ramosum	1	1	1	1	1	1	1	1	8	2.08	16.64
Ceratonia siliquosa	1	0	0	1	0	0	0	0	2	0.69	1.39
Chamaerops humilis	1	0	1	0	0	0	0	1	3	1.10	3.30
Cortaderia selloana	0	1	0	0	1	1	1	1	5	1.61	8.05
Erica multiflora	1	0	1	1	0	1	0	0	4	1.39	5.55
Euphrasia sp.	1	1	1	0	0	1	1	1	6	1.79	10.75
Lavandula augustifolia	1	0	1	1	0	0	0	0	3	1.10	3.30
Phillyrea augustifolia	1	1	1	1	1	1	1	1	8	2.08	16.64
Pinus halepensis	1	1	1	1	1	1	1	1	8	2.08	16.64
Pistacea lentiscus	1	0	1	1	0	1	0	0	4	1.39	5.55
Quercus coccifera	1	1	1	1	1	1	1	1	8	2.08	16.64
Rosmarinus officinalis	1	1	1	1	0	1	1	1	7	1.95	13.62
Salvia sp.	1	0	1	1	0	0	0	0	3	1.10	3.30
Sedum sp.	0	1	0	0	1	0	1	1	4	1.39	5.55
Smilax asper	0	0	0	0	0	0	0	1	1	0	0
Total species number, α	**12**	**8**	**11**	**10**	**6**	**9**	**8**	**10**	**74**		**Sum = 126.9**
$\ln \alpha$	2.48	2.08	2.40	2.30	1.79	2.20	2.08	2.30			
$\alpha \times \ln \alpha$	29.82	16.64	26.38	23.03	10.75	19.78	16.64	23.03	**Sum = 166**		

1, species present; 0, species not present; q, quadrat.

Wilson and Schmida's β_T

This is calculated using the parameters calculated in Whittaker's β_w and Cody's β_c above: $\beta_T = (5 + 3)/(2 \times 9.5) = 0.421$.

Routledge's β_R

The total species number for the transect, $S = 15$. To calculate the total number of species with overlapping distributions form a species by species matrix and mark on it all species pairs that occur in at least one quadrat. For our data set *Brachypodium ramosum* was found in every quadrat so it must overlap with every other species in at least one quadrat.

Species	1	2	3	4	5	6	7	8	9	10	11	12	13	14	15	*N*
1		×	×	×	×	×	×	×	×	×	×	×	×	×	×	14
2			×		×	×	×	×	×	×	×	×	×			10
3				×	×	×	×	×	×	×	×	×	×	×	×	12
4					×	×	×		×	×	×	×		×	×	9
5						×	×	×	×	×	×	×	×			8
6							×	×	×	×	×	×	×	×	×	9
7								×	×	×	×	×	×			6
8									×	×	×	×	×	×	×	7
9										×	×	×	×	×	×	6
10											×	×	×			3
11												×	×	×	×	4
12													×	×	×	3
13																
14															×	1
15																
Total number of joint occurrences																92

For this data set there are 92 joint occurrences, r, so that $\beta_R = 15 \times 15/((2 \times 92) + 15) - 1 = 225/199 - 1 = 0.13$.

Routledge's β_I and β_E

The Occurrence column in Table 11.1 gives the number of quadrats in which each species was recorded, e_i. The sum of these values, $T_i = 74$. The calculations use natural (base e) logarithms. First calculate $\Sigma e_i \ln e_i = 126.9$ and $\Sigma \alpha_i \ln \alpha_i = 166$ as shown in Table 11.1. Placing these values in the formula gives $\beta_I = \ln 74 - ((1/74) \times 126.9) - ((1/74) \times 166) = 4.3 - 1.71 - 2.24 = 0.35$, and β_E is then easily calculated as $\exp(\beta_I) = 1.42$.

Alternative approaches for beta diversity

A different approach to β diversity is to measure α diversity along a recognized gradient or at least a linear transect. The slope of the line will measure β diversity and sudden changes will reflect community or higher order boundaries (Odum *et al.* 1960). Pattern analysis, much used in plant ecology (Kershaw 1973; Goldsmith & Harrison 1976; Greig-Smith 1978), adopts this approach on a small scale.

Wilson and Mohler (1983) discuss the calculation of β diversity for quantitative data. A direct approach to β diversity would be to compare the separate diversity indices, and the Shannon index (see page 123) has sometimes been used, but given the shortcomings of this index as a measure of α diversity it seems inappropriate as a general technique. The most effective approach is based on the use of similarity indices. There is a great number of similarity measures that might be used; generally the best index for presence-absence data is the Jaccard and for quantitative data the Morisita–Horn (Wolda 1981, 1983; Magurran 1988). Smith (1986) also concluded that for quantitative data the Morisita–Horn index was one of the most satisfactory. These similarity indices are described in Chapter 12.

ea Example applications

β diversity calculations will normally be undertaken on data collected along a transect. Below are given some typical examples of sampling along a gradient that will yield β diversity measures.

Transects down a beach

The duration of exposure at low tide is an important variable determining the species fauna of soft sediments on beaches. To measure the rate of turnover of species establish a transect down the beach at low water and take core samples (see page 26)

(continued on page 134)

ea Example applications *(continued)*

at regular intervals along the transect. An example of data obtained in this manner is given in Table 11.2. Using Whittaker's β_w, the β diversity of each transect was calculated as follows:

Transect 1:
Total species recorded = 19
Mean species richness (α) = 9
β diversity = (19/9) − 1 = 1.11

Transect 2:
Total species recorded = 16
Mean species richness (α) = 7.75
β diversity = (16/7.75) − 1 = 1.06

Table 11.2 Results of an intertidal survey in Southampton Water. Two transects were established down the beach and four core samples taken at 10 m intervals along each transect.

	T1 S1	T1 S2	T1 S3	T1 S4	T2 S1	T2 S2	T2 S3	T2 S4
Cerastoderma edule	—	—	1	2	—	1	—	1
Abra nitida	1	—	—	—	—	—	1	—
Abra alba	—	—	—	—	—	—	3	1
Abra sp.	—	1	3	1	—	—	—	—
Littorina littorea	—	—	1	—	2	—	8	1
Hydrobia ulvae	23	36	16	3	9	127	77	7
Carcinus maenas	—	—	—	—	—	—	1	1
Cyathura carinata	6	38	2	—	—	—	2	3
Microdentopterus sp.	—	1	1	—	—	—	—	—
Amphipoda indet.	—	3	—	—	—	—	—	—
Amphilochus cf. *neapolitanus*	—	3	1	1	—	1	—	—
Corophium volutator	—	—	—	1	—	—	—	—
Corophium sp.	—	—	—	—	—	1	—	—
Amphipoda indet.	—	—	—	—	—	—	1	—
Jaera albifrons grp	13	55	287	123	180	144	246	192
Tubificoides	3	—	—	—	—	—	—	—
Hediste diversicolor	5	—	2	—	15	3	1	6
Nephtys hombergi	—	—	—	1	—	—	1	—
Sabellid indet.	—	10	1	—	—	—	—	—
Monayunkia aestuarina	—	1	—	—	—	—	—	—
Amphorete acutifrons	—	—	2	—	—	—	—	—
Cirriformia tenticulata	—	—	—	1	—	—	—	8
Polydora sp.	—	—	—	—	1	—	—	—
Anemones	—	1	4	—	—	—	1	—
Species number	6	10	12	8	5	6	11	9

S, station; T, transect.

(continued)

ea

Example applications *(continued)*

The β diversity of both transects is almost identical, suggesting that the species turnover down the beach was the same along both transects.

Transects up a hillside

Vegetation can change rapidly moving from a valley up the slope of a hill. The rate of species turnover can be calculated and compared at a number of localities by establishing transects and quadrat sampling at set distances (or altitudes) along the transect. The calculations undertaken could be identical to those used above or methods based on quantitative data could be used.

Salinity gradients

Estuaries show dramatic salinity gradients from freshwater in the river to full strength seawater at the mouth. Almost any animal or plant group can be sampled at various localities along the estuary and the diversity calculated. You should not assume that salinity follows a simple gradient with distance from the sea. If salinity data is unavailable, you should measure the salinity at each sampling locality at different states of the tide.

Exposure on rocky shores

In most coastal areas there is a prevailing wind direction and some parts of the coast are more exposed to wave action than others. Sampling along a transect across a zone that ranges from more to less exposed will generate data that can be used to explore β diversity.

Flow rates in streams

The fauna of a stream changes with the flow regime so that sheltered localities with reduced flow hold different species of plant, insect, and fish. By sampling along a transect moving from high to low flow conditions the effect of flow on diversity can be measured. There are clear practical applications to such studies, as recently there have been claims that there is a lack of habitat diversity in streams due to man removing debris and simplifying stream architecture so that there is reduced variation in flow. Such studies can be used to estimate increase in diversity that could be produced by habitat enhancement.

Chapter Twelve

Comparing and classifying communities

Even a quite modest ecological survey can produce a bewildering amount of information on the presence and abundance of species and it is frequently difficult to identify and summarize the main features and interrelationships between communities. This chapter describes the different approaches and techniques you can use. The aim is to describe the ways in which your data can be presented and explored and little attention is given to the theoretical background of the techniques. You will not be able to carry out any of these techniques without a computer and specialist statistical or ecological software. Searching the internet using the names of the techniques will give addresses where software can be purchased or obtained for free. All of the examples in this chapter were calculated using the COMMUNITY ANALYSIS PACKAGE (CAP) produced by Pisces Conservation Ltd (www.irchouse.demon.co.uk). Useful sources for information about the mathematical background to these techniques are Legendre and Legendre (1998) and Kent and Coker (1992).

Organizing the data for analysis

Ecological data sets can often be arranged as a two-dimensional matrix consisting of n samples (or stations) forming the columns and the S species forming the rows. This matrix or grid can hold either the observed abundance of each species, an abundance score, or presence-absence information. Presence-absence data is just recorded as 1 if it was found in the sample and 0 if not. The most convenient way to store and organize these data is with the use of a spreadsheet program such as EXCEL or LOTUS 1-2-3. Most statistical and multivariate software can import data from these spreadsheets. If the objective is to search for species interrelationships then the S column by S row matrix of species correlations or similarities may be formed and what is termed an R analysis undertaken. If the objective is to identify samples with similar communities then the n column by n row matrix of sample correlations or similarities may be used in a Q analysis.

Searching for similarity

When we compare the flora or fauna sampled at different localities we often wish to know how similar are their species assemblages. Numerous methods have been devised for the measurement of similarity, the most successful of which are described below. Legendre and Legendre (1998) give a more complete account of similarity and distance measures.

Similarity indices are simple measures of the extent to which either two habitats have species in common (Q analysis) or species have habitats in common (R analysis). Binary similarity coefficients use presence-absence data and more complex quantitative coefficients can be used if you have data on species abundance. When comparing the species at two localities indices can be divided into those that take account of the absence of a species from both communities (double zero methods) and those that do not. In most ecological applications it is unwise to use double zero methods as they assign a high level of similarity to localities that **both** lack many species. We would not normally consider two sites highly similar because their only common feature was the joint lack of a group of species, which could occur because of sampling errors or because both sites were unsuitable, so here are only described binary indices that exclude double zeros.

Binary coefficients for presence-absence data

When comparing two sites let a be the number of species held in common and b and c the number of species found at only one of the sites. When comparing two species over many sites the terms are similar, e.g. a is the number of sites where they both were caught. The three simplest coefficients are given below.

Jaccard:

$$C_j = \frac{a}{(a+b+c)} \qquad (12.1)$$

Sørensen:

$$C_s = \frac{2a}{(2a+b+c)} \qquad (12.2)$$

Mountford:

$$C_M = \frac{2a}{2bc-(b+c)a} \qquad (12.3)$$

The Mountford index was designed to be less sensitive to sample size than either the Sørensen or Jaccard; however, it assumes that species abundance fits a log-series model which may be inappropriate.

Following evaluation of similarity measures using Rothamsted insect data, Smith (1986) concluded that no index based on presence-absence data was entirely satisfactory, but the Sørensen was the best of those considered.

Quantitative coefficients

Based purely on species number, binary coefficients give equal weight to all species and hence tend to place too much significance on the rare species whose capture will depend heavily on chance. Bray and Curtis (1957) brought abundance into consideration in a modified Sørensen coefficient and this approach is widely used in plant ecology; the coefficient as modified essentially reflects the similarity in individuals between the habitats:

$$C_N = \frac{2jN}{(aN + bN)} \tag{12.4}$$

where aN = the total individuals sampled in habitat a, bN = the same in habitat b, and jN = the sum of the lesser values of abundance in both samples for the species common to both habitats (often termed W).

However, for quantitative data Wolda (1981, 1983) found that the only index not strongly influenced by sample size and species richness was the Morisita–Horn index:

$$C_{MH} = \frac{2\sum(an_i \cdot bn_i)}{(\mathrm{d}a + \mathrm{d}b)aN \cdot bN} \tag{12.5}$$

where aN and bN = the total number of individuals in sites a and b respectively, an_i and bn_i = the number of individuals in the ith species in sample a and b respectively, and

$$\mathrm{d}a = \frac{\sum an_i^2}{aN^2} \tag{12.6}$$

db is calculated in similar fashion using sample b.

A considerably simpler but frequently used index is percent similarity (Whittaker 1952), calculated using:

$$P = 100 - 0.5\sum_{i=1}^{S}|P_{a,i} - P_{b,i}| \tag{12.7}$$

where $P_{a,i}$ and $P_{b,i}$ are the percentage abundances of species i in samples a and b respectively and S is total species number. This index takes little account of rare species and thus will give a good indication of the similarity in dominant forms of life between the sites.

$$\sum_{i=1}^{S}|P_{a,i} - P_{b,i}|$$

Multivariate analysis

Multivariate analysis is used when the objective is to search for relationships between or classify objects (sites or species) that are defined by a number of attributes. Generally, we seek to show the relationship between sites (or samples) using the species as the attributes. Data sets can be large, e.g. marine benthic or forest beetle faunal studies can easily require analysis of a matrix of 100 samples (stations) by 350 species and thus multivariate analysis requires a computer. If the objective is to assign objects to a number of discrete groups then cluster analysis should be considered. If there is no a priori reason to believe the objects will or could naturally fall into groups, then an ordination technique may be more suitable. Ordination assumes the objects form a continuum of variation and the objective is often to generate hypotheses about the environmental factor(s) that mould community structure.

There is a considerable literature on multivariate techniques. Useful texts for ecologists are Legendre and Legendre (1998), Digby and Kempton (1987), and Kent and Coker (1992).

Cluster analysis

When numbers of sites or habitats are to be compared the similarity measures described above can form the basis of cluster analysis, which seeks to identify groups of sites, or stations, that are similar in their species composition.

Classification methods comprise two principal types, hierarchical, where objects are assigned to groups that are themselves arranged into groups as in a dendrogram, and nonhierarchical, where the objects are simply assigned to groups. The methods are further classified as either agglomerative where the analysis proceeds from the objects by sequentially uniting them or divisive, where all the objects start as members of a single group that is repeatedly divided. For computational and presentational reasons hierarchical-agglomerative methods are the most popular.

The basic computational scheme used in cluster analysis can be illustrated using single linkage cluster analysis as an example. This is the simplest procedure and consists of the following steps:

1 Start with n groups each containing a single object (sites or species).

2 Calculate, using the similarity measure of choice, the array of between-object similarities.

3 Find the two objects with the greatest similarity, and group them into a single object.

4 Assign similarities between this group and each of the other objects using the rule that the new similarity will be the greater of the two similarities prior to the join.

5 Continue steps 3 and 4 until only one object is left.

The results from a cluster analysis are usually presented in the form of a dendrogram. The dendrogram in Figure 12.1 was undertaken on the dune plant data given in Table 12.1 and shows clearly that sites 5, 6, 7, and 10 form a tight group with considerable similarities in their species composition. At the opposite extreme, site 17 has little in common with these sites as can be confirmed from an examination of Table 12.1.

The problem with all classification methods is that there can be no objective criteria of the best classification; indeed even randomly generated data can produce a pleasing dendrogram. Always consider carefully if the

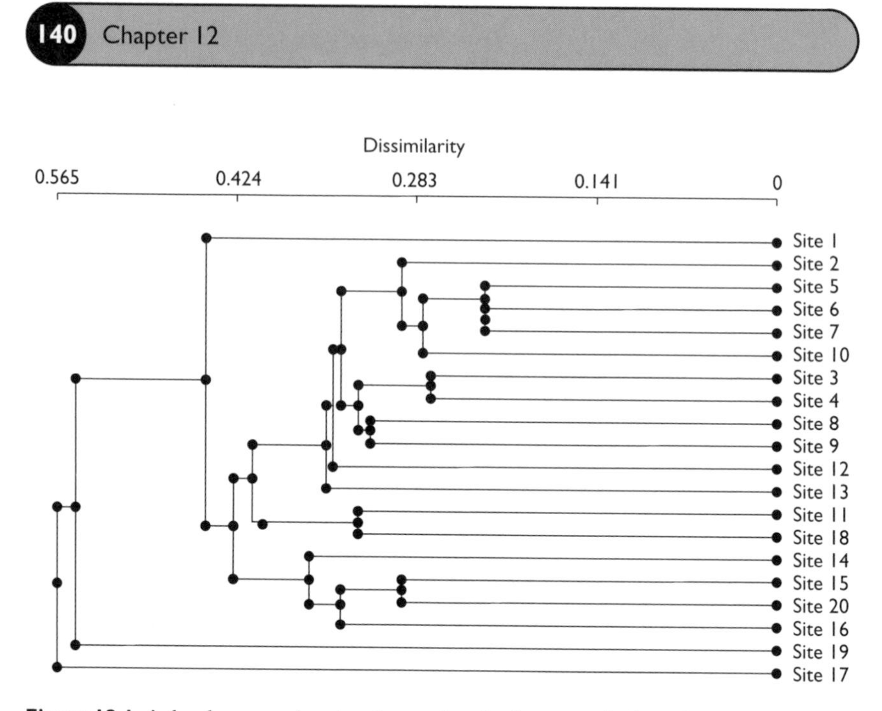

Figure 12.1 A dendrogram showing the results of a cluster analysis undertaken on the plant data given in Table 12.1. The single linkage clustering method was used with the Bray–Curtis similarity measure. (Graphical output produced by COMMUNITY ANALYSIS PACKAGE, www.irchouse.demon.co.uk.)

groupings identified seem to make sense and reflect some feature of the natural world.

TWINSPAN

This method can be used with presence-absence, percentage cover and quantitative data. The ability to effectively handle percentage cover makes the method attractive to botanists. The Two-Way INdicator SPecies ANalysis procedure (Hill *et al.* 1975) also produces dendrograms of the relationship between species and samples but uses the reciprocal averaging ordination method to order the species and samples. Thus the method is something of a hybrid between classificatory and ordination methods. It is particularly attractive in studies where the objective is to classify communities, so that field workers can quickly assign an area to a community type, and is much favored by botanists. This is because it identifies indicator species characteristic of each group identified. The output from TWINSPAN applied to the dune data (Table 12.1) is shown in Figure 12.2. At the left of the dendrogram it can be seen that the first partition in the sites had the indicator species Ranflo and Agrsto. Further, by comparing Figures 12.1 and 12.2, it can be seen that this method also shows the close similarity in sites 5, 6, 7, and 10.

TWINSPAN is a useful technique when you are seeking to identify species that can be used to characterize particular communities. It is, however, not

Table 12.1 An example of biological data: the dune meadow data used by Jongman *et al.* (1995). The latin names of the species have been abbreviated.

	Site																			
Species	**1**	**2**	**3**	**4**	**5**	**6**	**7**	**8**	**9**	**10**	**11**	**12**	**13**	**14**	**15**	**16**	**17**	**18**	**19**	**20**
Achmil	1	3	0	0	2	2	2	0	0	4	0	0	0	0	0	0	2	0	0	0
Agrsto	0	0	4	8	0	0	0	4	3	0	0	4	5	4	4	7	0	0	0	5
Airpra	0	0	0	0	0	0	0	0	0	0	0	0	0	0	0	0	2	0	3	0
Alogen	0	2	7	2	0	0	0	5	3	0	0	8	5	0	0	4	0	0	0	0
Antodo	0	0	0	0	4	3	2	0	0	4	0	0	0	0	0	0	4	0	4	0
Belper	0	3	2	2	2	0	0	0	0	2	0	0	0	0	0	0	0	2	0	0
Brohor	0	4	0	3	2	0	2	0	0	4	0	0	0	0	0	0	0	0	0	0
Chealb	0	0	0	0	0	0	0	0	0	0	0	0	1	0	0	0	0	0	0	0
Cirarv	0	0	0	2	0	0	0	0	0	0	0	0	0	0	0	0	0	0	0	0
Elepal	0	0	0	0	0	0	0	4	0	0	0	0	0	4	5	8	0	0	0	4
Elyrep	4	4	4	4	4	0	0	0	6	0	0	0	0	0	0	0	0	0	0	0
Empnig	0	0	0	0	0	0	0	0	0	0	0	0	0	0	0	0	0	0	2	0
Hyprad	0	0	0	0	0	0	0	0	0	0	2	0	0	0	0	0	2	0	5	0
Junart	0	0	0	0	0	0	0	4	4	0	0	0	0	0	3	3	0	0	0	4
Junbuf	0	0	0	0	0	0	2	0	4	0	0	4	3	0	0	0	0	0	0	0
Leoaut	0	5	2	2	3	3	3	3	2	3	5	2	2	2	2	0	2	5	6	2
Lolper	7	5	6	5	2	6	6	4	2	6	7	0	0	0	0	0	0	2	0	0
Plalan	0	0	0	0	5	5	5	0	0	3	3	0	0	0	0	0	2	3	0	0
Poapra	4	4	5	4	2	3	4	4	4	4	4	0	2	0	0	0	1	3	0	0
Poatri	2	7	6	5	6	4	5	4	5	4	0	4	9	0	0	2	0	0	0	0
Potpal	0	0	0	0	0	0	0	0	0	0	0	0	0	2	2	0	0	0	0	0
Ranfla	0	0	0	0	0	0	0	2	0	0	0	0	2	2	2	2	0	0	0	4
Rumace	0	0	0	0	5	6	3	0	2	0	0	2	0	0	0	0	0	0	0	0
Sagpro	0	0	0	5	0	0	0	2	2	0	2	4	2	0	0	0	0	0	3	0
Salrep	0	0	0	0	0	0	0	0	0	0	0	0	0	0	0	0	0	3	3	5
Tripra	0	0	0	0	2	5	2	0	0	0	0	0	0	0	0	0	0	0	0	0
Trirep	0	5	2	1	2	5	2	2	3	6	3	3	2	6	1	0	0	2	2	0
Viclat	0	0	0	0	0	0	0	0	0	1	2	0	0	0	0	0	0	1	0	0
Brarut	0	0	2	2	2	6	2	2	2	2	4	4	0	0	4	4	0	6	3	4
Calcus	0	0	0	0	0	0	0	0	0	0	0	0	0	4	0	3	0	0	0	3

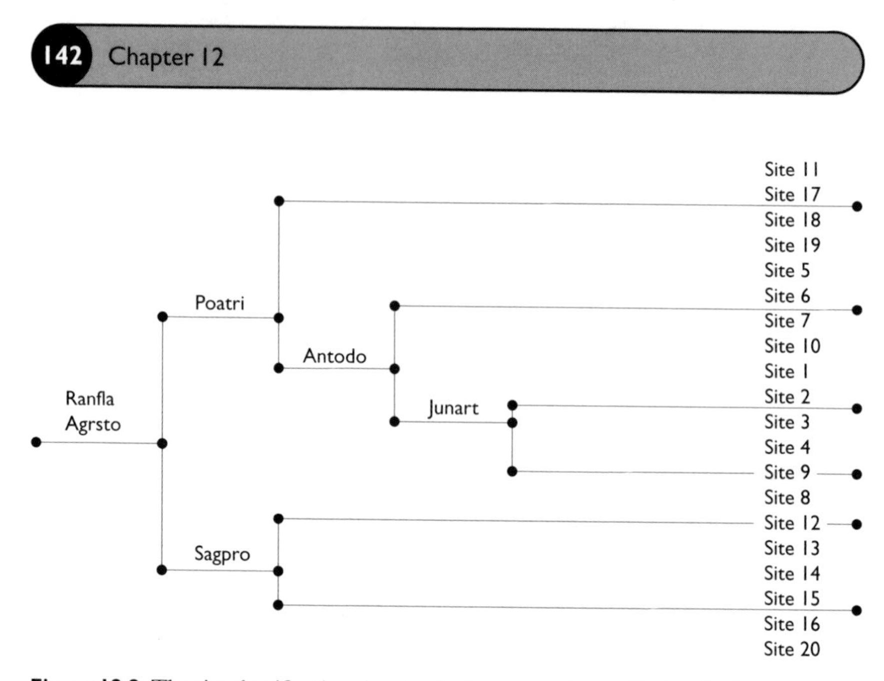

Figure 12.2 The site classification showing indicator species applied to the dune plant data given in Table 12.1. (Graphical output produced by COMMUNITY ANALYSIS PACKAGE, www.irchouse.demon.co.uk.)

always an easy method to understand. A particular oddity of the method is the concept of "pseudospecies". Each species is divided into a number of pseudospecies which represent the different abundance levels at which it was found. To undertake a TWINSPAN analysis you will need to obtain specialist software (e.g. CAP produced by Pisces Conservation Ltd, www.irchouse.demon.co.uk).

Ordination

A number of ordination techniques are commonly used by ecologists and it is not possible to give clear guidance as to the method that should be used. In community studies the aim of the analysis is to summarize the relationship between the samples (sites) and thus the best method is the one that gives the clearest and most interpretable picture. Therefore it may be appropriate to try a number of methods and compare the results.

Principal Components Analysis (PCA)

PCA is the oldest and still one of the most frequently used ordination techniques in community ecology. It is most appropriate for full quantitative data, but can be used if abundance is classified into a number of abundance classes. The objective of the method is to express the relationship between the samples in a two- or three-dimensional space that can be plotted and usefully visualized. This can only be achieved if many of the species are positively or negatively correlated. Normally this will be so for a number of reasons.

Firstly, there is interdependence between organisms in an ecosystem and secondly, many species respond similarly to environmental variables such as temperature and water. General descriptions of the procedure for biologists are given (Digby & Kempton 1987; Kent & Coker 1992; Legendre & Legendre 1998).

The analysis is undertaken on either the between-site variance–covariance or correlation matrix. If the species vary greatly in abundance you will probably need to transform the data by taking logarithms or using a square root transformation. Logarithmic transformations would be excellent if it were not for the fact that zeros cannot be handled. A frequently used procedure is to add 1 to all the observations. This can distort the output and it is probably more appropriate to use a square root transformation. If you undertake a PCA on the correlation matrix you will be giving all species, irrespective of abundance, equal weighting, whereas the analysis undertaken on the variance–covariance matrix will reflect differences in abundance, but can result in the dominant species determining the output. When successful, PCA will present major features of a complex ecological community in only two or three dimensions and the ordination of sites along these new axes can be related to underlying environmental factors that are molding community structure. PCA can be judged a success when the first two or three principal axes explain an appreciable proportion of the total variability in the data set. For large ecological data sets with more than 20 species if the three largest axes can explain more than 30% of the variance this would be good.

The output from a PCA applied to the dune data (Table 12.1) is shown in Figure 12.3. At the left of the plot of the first two principal axes it can be seen

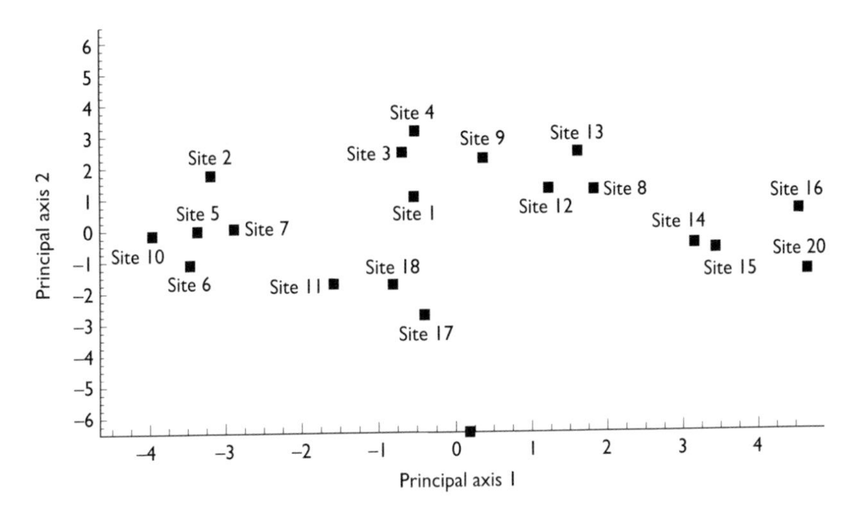

Figure 12.3 The results of principal components analysis to produce an ordination of the dune data given in Table 12.1. (Graphical output produced by COMMUNITY ANALYSIS PACKAGE, www.irchouse.demon.co.uk.)

that sites 5, 6, 7, and 10 are again clustered together as they were in Figures 12.1 and 12.2.

Nonmetric Multidimensional Scaling (NMDS)

NMDS can be viewed as a more generalized development of PCA. Benthic ecologists working with full quantitative data from grab samples particularly favor this method. It is widely applicable and frequently produces a better ordination than PCA. It takes more computational effort than PCA and has a potentially important feature that should be noted. There is no guarantee that the ordination will be the best possible and it is therefore essential to do a number of runs with different starting conditions to be sure that you have obtained a stable ordination.

The output from an NMDS ordination applied to the dune data (Table 12.1) is shown in Figure 12.4. At the left of the plot it can be seen that sites 5, 6, 7, and 10 are again clustered together as they were in Figures 12.1–12.3.

Reciprocal Averaging (RA)

RA (also termed correspondence analysis) and an adjusted version called DEtrended CORrespondence ANAlysis (DECORANA) are the final types of ordination method that are frequently used. They are best used on quantitative data although they can give good results with classed abundance data. Both of these methods are particularly effective when it is suspected that the sites can be arranged along an environmental gradient. Further, the method allows the site and species ordinations to be plotted on the same figure, which

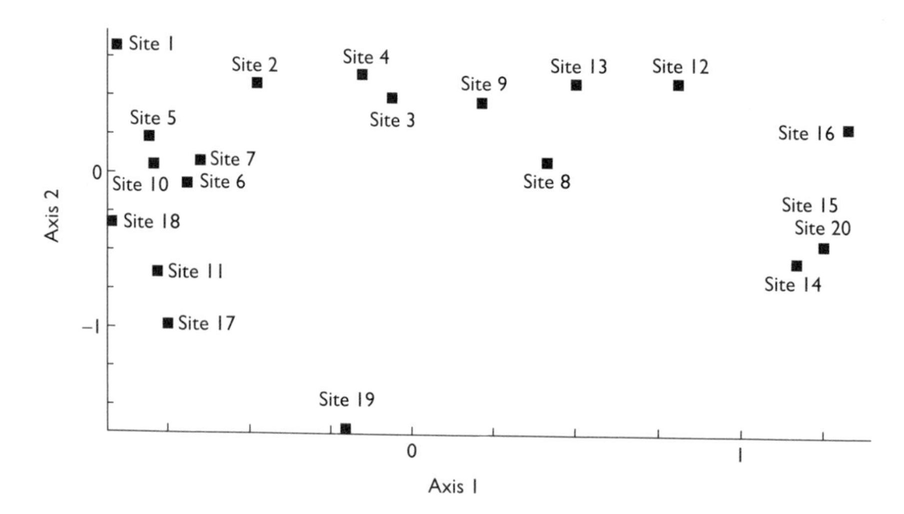

Figure 12.4 The results of an ordination of the dune data given in Table 12.1 using non-metric multidimensional scaling. (Graphical output produced by COMMUNITY ANALYSIS PACKAGE, www.irchouse.demon.co.uk.)

allows the influence of the species in determining the ordination of the sites to be uncovered. The output from an RA ordination applied to the dune data (Table 12.1) is shown in Figure 12.5. It can be seen that sites 5, 6, 7, and 10 are again clustered together as they were in all the previous methods (Figs 12.1–12.4). However in general the site ordination is not quite so well spread across the graph. As with all the previous methods sites 17 and 19 are clearly different from the others. Examination of the species ordination (Fig. 12.5) shows that this is related to the presence of the species labeled Empnig, Hyprad, and Airpra, a fact that is confirmed from an inspection of Table 12.1. The ability of the method to associate species with particular clusters of sites can be of great value.

Identifying influential environmental variables

Assuming that in addition to collecting the species abundance data, information on the environment at each sampling site was also collected, it is possible to investigate the relationship between the species assemblages and the environment. Ordination methods such as PCA and RA can suggest which environmental variables are most influential by examining the site ordination. The simplest way is to plot out the ordination and write next to each site the value for a particular environmental variable and then see if it forms a

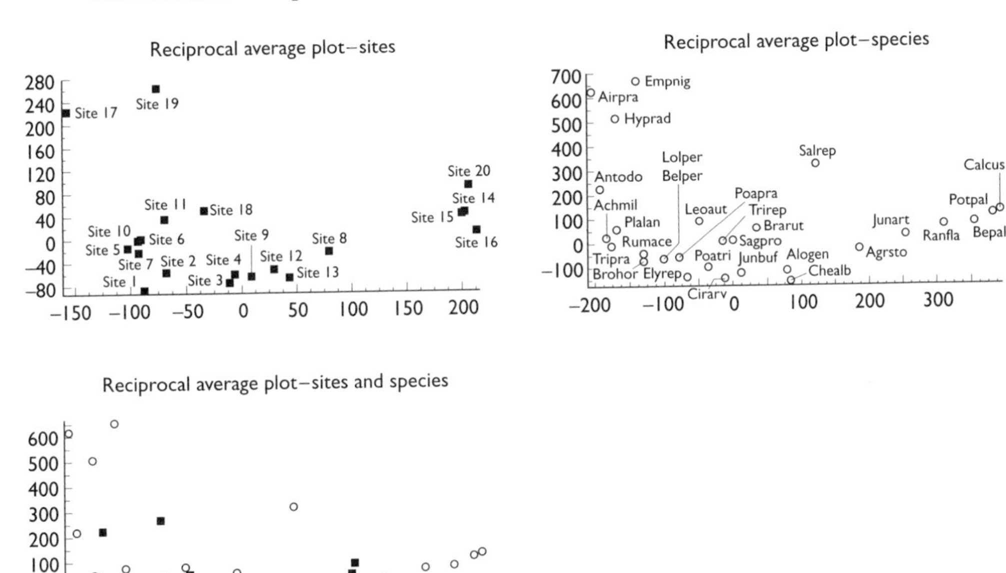

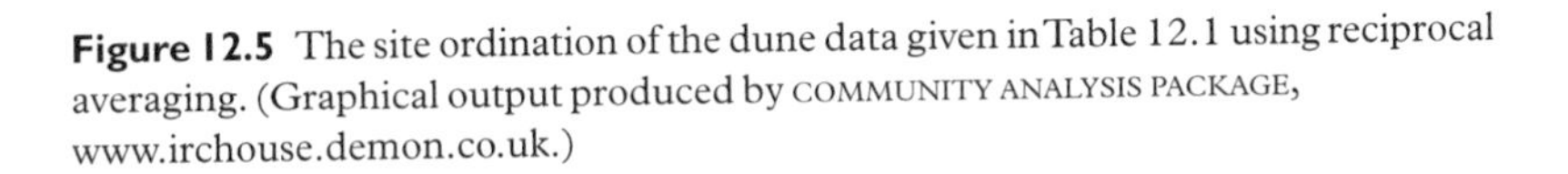

Figure 12.5 The site ordination of the dune data given in Table 12.1 using reciprocal averaging. (Graphical output produced by COMMUNITY ANALYSIS PACKAGE, www.irchouse.demon.co.uk.)

gradation in value along any axis. For example, if in Figure 12.3 the sites at the left hand side of axis 1, sites 10, 5, 6, and 7, were all from low pH soils, and the sites at the right hand side of the axis, sites 15, 20, 16, and 14, were all from high pH soils, we might consider that axis 1 was a soil acidity axis. This rather ad hoc approach lacks rigor and cannot be used to explore the common situation where two or more environmental variables are interacting to produce the suite of conditions required by the different species. If you have sufficiently good data it is possible to use a superior approach called Canonical Correlation Analysis (CCA).

To undertake a CCA you require data on the environmental conditions at each site. These data can be classificatory variables such as exposed (=1) and sheltered (=0) or fully quantitative readings such as salinity or temperature. As an example, Table 12.2 gives environmental data for the dune sites presented in Table 12.1. The output from CCA is extensive and beyond the scope of the present work. The analysis will indicate if one or more of the environmental variables are correlated with the observed species assemblages. If a significant influence is identified, perhaps the most useful part of the output is the graphs that show the organization of the species along a particular environmental gradient. An example of this plot for the dune data is shown in Figure 12.6.

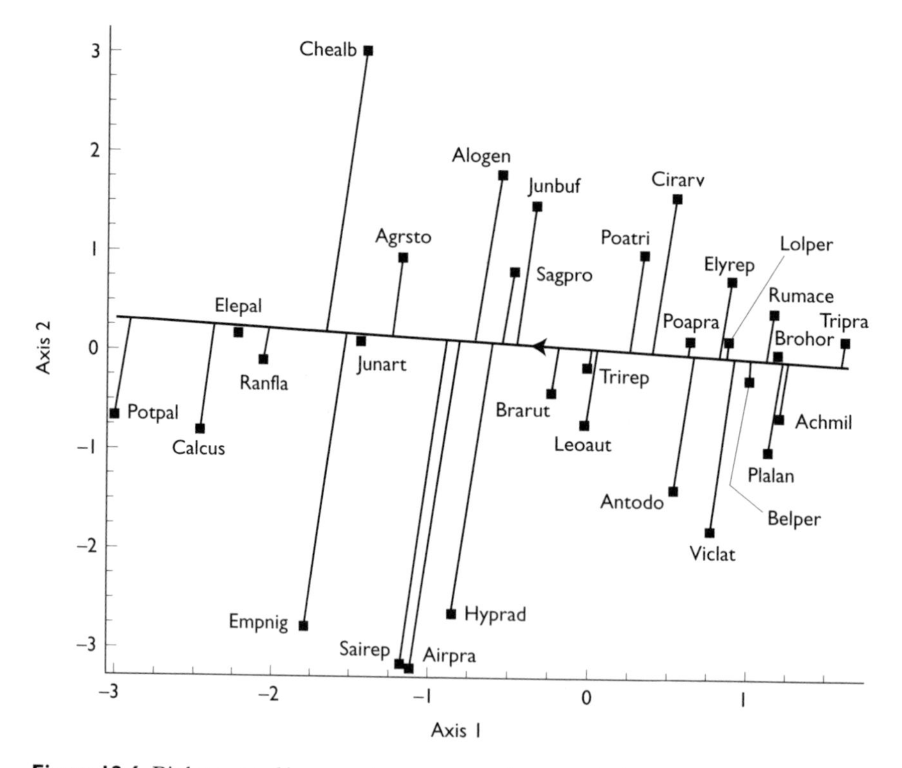

Figure 12.6 Biplot created by canonical correspondence analysis. In this plot the species are arranged along the vector representing the depth of the A1 soil horizon. Species to the left favor deeper soil. (Graphical output produced by ECOM, www.irchouse.demon.co.uk.)

Table 12.2 Environmental data for the dune sites for which the species data is given in Table 12.1. (From Jongman *et al.* 1995.)

	Site																			
	1	2	3	4	5	6	7	8	9	10	11	12	13	14	15	16	17	18	19	20
A1	2.8	3.5	4.3	4.2	6.3	4.3	2.8	4.2	3.7	3.3	3.5	5.8	6	9.3	11.5	5.7	4	4.6	3.7	3.5
Moisture	1	1	2	2	1	1	1	5	4	2	1	4	5	5	5	5	2	1	5	5
Manure	4	2	4	4	2	2	3	3	1	1	1	2	3	0	0	3	0	0	0	0
Hayfield	0	0	0	0	1	0	0	0	1	1	0	0	0	0	0	0	1	1	1	1
Pasture	0	0	0	0	0	0	1	1	0	0	1	0	0	1	0	1	0	0	0	0
SF	1	0	1	1	0	0	0	0	0	0	0	1	1	0	0	1	0	0	0	0
BF	0	1	0	0	0	0	0	0	0	1	1	0	0	0	0	0	0	0	0	0
NM	0	0	0	0	0	0	0	0	0	0	0	0	0	1	1	0	1	1	1	1

A1, the thickness of the A1 soil horizon; Moisture, the moisture content of the soil; Manure, the amount of manuring in five classes; Hayfield, the grassland management type; Pasture, the grassland use; SF, standard farming methods; BF, biological farming methods; NM, nature conservation management.

The gradation of species abundance with respect to soil depth is clearly presented. Using this diagram it is immediately possible to identify those species that favor a deeper A1 soil horizon, as they are at the left hand of the line of the vector that represents the depth of the soil.

CCA will only yield useful results if the key environmental variables influencing species distribution have been measured. Further, the method is sensitive to the quality of the data and poor quality or incomplete data will likely produce little of value while giving an impressively scientific output.

Measurement of interspecific association

These measurements may be based on either presence-absence or quantitative abundance data; as Hurlbert (1971) points out, presence-absence data is preferable if it is desired to measure the extent to which two species' requirements are similar. Interspecific competition may lead to a "misleading" lack of association if the measure is based on abundance. However it would seem that whenever possible both types of analysis should be undertaken, for a positive association on presence-absence data and a much weaker or negative one on abundance data would suggest (not prove) interspecific competition worthy of further analysis.

The departure of the distribution of presence-absence from independence

These methods measure the departure from independence of the distribution of the two species and assume that the probability of occurrence of the species is constant for all samples. Thus if the distribution of two aphid predators were being compared it would only be legitimate to include samples that also contained the prey. A good example of the use of these indices is given by Evans and Freeman (1950), who measured the interspecific association of two species of flea on two different rodents (*Apodemus* and *Clethrionomys*); they found that there was a strong negative association on *Apodemus* and a moderate positive one on *Clethrionomys* and suggested that the coarse and longer fur of *Clethrionomys* allowed the two fleas to avoid competition and exist together on that host.

When applied to habitats whose uniformity is doubtful, these methods give results of little value.

If two species are rare and therefore both are absent from most of the samples (this could, as just indicated, be due to unsuitable samples being included), a high level of association will be found. Conversely, if two species occur in most of the samples and so are nearly always found together, no association will be shown with these methods. (Although a biological association is obvious, it is equally correct to say that this does not depart significantly from an association that is due to chance and therefore does not necessarily imply any interspecific relationship.)

The two-dimensional Kolmogorov–Smirnov test

Garvey *et al.* (1998) proposed that a statistical technique used by astronomers could be used to detect the association between two species. The observations for each species are plotted as a scatter plot, each pair of coordinates (X,Y) being taken in turn and the number of points in each of the four surrounding quadrats (with (X,Y) as the origin) counted. The counts for the four quadrats are expressed as proportions. These proportions are compared with those that would be expected if the two species were independent. The expected values are calculated by multiplying the counts of the proportions of observations $<X$ and $\geq X$ and $<Y$ and $\geq Y$. Within each quadrat the difference between the observed and expected proportion of points is determined and the maximum difference between the observed and expected over all the quadrats and points found. The original pairs of data points are then randomized 5000 times using a computer and the test calculations preformed on each randomization. The statistical difference is then determined by comparing the observed maximum difference against the randomly created distribution of maximum differences.

The contingency table

The basis of these methods is the 2×2 contingency table.

	Species A		
	Present	**Absent**	**Total**
Species B			
Present	a	b	$a+b$
Absent	c	d	$c+d$
Total	$a+c$	$b+d$	$n=a+b+c+d$

Such a table should always be drawn up so that A is more abundant than B, i.e. $(a+b) < (a+c)$. A number of statistics are available for analysis of such a table, but the corrected chi-square (χ^2) makes fewest assumptions about the type of distribution, and the significance of the value obtained can be determined from tables available in standard statistical textbooks. It is calculated using:

$$\chi^2 = \frac{n[|ad - bc| - (n/2)]^2}{(a+c)(b+d)(a+b)(c+d)} \tag{12.9}$$

The test in this form is only valid if the expected numbers (if the distribution was random) are not less than 5; there is only one degree of freedom and so the 5% point is 3.84. Therefore, if a χ^2 of less than this is obtained, any apparent association could well be due to chance and further analysis should be abandoned. If the smallest expected number is less than 5 the exact test should be used.

Coefficients of association

If χ^2 is significant then one of the coefficients of association may be used to give an actual quantitative value for comparison with other species. They are designed so that the coefficient has the same range as the correlation coefficient (r), i.e. $+1$ = complete positive association, -1 = complete negative association, and 0 = no association. Cole (1949) reviews a number of coefficients and points out how, if the above interpretations of values from $+1$ to -1 are to hold for comparative purposes, the plot of the value of the coefficient against the possible number of joint occurrences should be linear; for several of the coefficients it is not, and for those in which it is, the plot does not pass through zero. Some coefficients are given below.

Coefficient of mean square contingency

This coefficient makes no assumption about distribution, but it cannot give a value of $+1$ unless $a = d$ and b and $c = 0$. For less extreme forms of association it is useful and easily calculated:

$$C_{AB} = \sqrt{\frac{\chi^2}{n + \chi^2}} \tag{12.10}$$

where C_{AB} = coefficient of association between species A and species B, n = total number of occurrences, and the χ^2 value is obtained as above.

Coefficient of interspecific association

The coefficients originally designed by Cole (1949) have been shown by Hurlbert (1969) to be biased by the species frequencies. He showed this bias was considerably diminished if the coefficient is:

$$C^1_{AB} = \frac{ad - bc}{\left| ad - bc \right|} \left| \left(\frac{Obs\chi^2 - Min\,\chi^2}{Max\,\chi^2 - Min\,\chi^2} \right)^{\frac{1}{2}} \right| \tag{12.11}$$

$Min\,\chi^2$ is the value of χ^2 when the observed a differs from its expected value ($\hat{a}$) by less than 1.0 (except when $a - \hat{a} = 0$ or $= 0.5$, the value of $Min\,\chi^2$ depends on whether $(ad - bc)$ is positive or negative), formulated:

$$Min\,\chi^2 = \frac{n^3(\hat{a} - g[\hat{a}])^2}{(a+b)(a+c)(c+d)(b+d)} \tag{12.12}$$

where $g(\hat{a}) = \hat{a}$, rounded to the next lowest integer when $ad < bc$ or rounded to the next highest integer when $ad > bc$, if $\hat{a}$ is an integer then $g(\hat{a}) = \hat{a}$.

$Max\,\chi^2$ is the value of χ^2 when a is as large or small as the marginal totals of the 2×2 table will permit formulated under specified conditions as detailed below.

$ad \geq bc$:

$$\chi^2 = \frac{(a+b)(b+d)n}{(a+c)(c+d)} \tag{12.13}$$

$ad < bc, a \leq d$:

$$\chi^2 = \frac{(a+b)(a+c)n}{(a+b)(c+d)} \tag{12.14}$$

$ad < bc, a > d$:

$$\chi^2 = \frac{(b+d)(c+d)n}{(a+b)(a+c)} \tag{12.15}$$

Obs χ^2 is calculated in the normal manner (see above). Pielou and Pielou (1968) describe a method for species of infrequent occurrence.

Proportion of individuals occurring together

Correlation may be utilized if the data can be normalized, otherwise some of the methods of comparing similarity between habitats may be used. For example Sørensen's coefficient, as used by Whittaker and Fairbanks (1958), may be modified to give the normal range of -1 (no association) to $+1$ (complete association):

$$I_{ai} = 2\left[\frac{\mathcal{J}}{A+B} - 0.5\right] \tag{12.16}$$

where $\mathcal{J}$ = number of individuals of A and B in samples where both species are present and A and B = total of individuals of A and B in all samples.

References

Adler, G. H. & Lambert, T. D. (1997) Ecological correlates of trap response of a Neotropical forest rodent, *Proechimys semispinosus*. *J. Trop. Ecol.* **13**, 59–68.

Allsteadt, J. & Vaughan, C. (1992) Population status of *Caiman crocodilus* (Crocodylia: Alligatoridae) in Cano Negro, Costa Rica. *Brenesia* **38**, 57–64.

Anwar, N. A., Richardson, C. A. & Seed, R. (1990) Age determination, growth rate and population structure of the horse mussel *Modiolus modiolus*. *J. Mar. Biol. Assoc. UK* **70**, 441–58.

Arnold, A. J. (1994) Insect suction sampling without nets, bags or filters. *Crop Protection* **13**, 73–6.

Bagenal, T. B. (1972) The variability in the number of perch, *Perca fluviatilis* L., caught in traps. *Freshwater Biol.* **2**, 27–36.

Bailey, N. T. J. (1951) On estimating the size of mobile populations from recapture data. *Biometrika* **38**, 293–306.

Bailey, N. T. J. (1952) Improvements in the interpretation of recapture data. *J. Anim. Ecol.* **21**, 120–7.

Baker, G. H. & Vogelzang, B. K. (1988) Life history, population dynamics and polymorphism of *Theba pisana* (Mollusca: Helicidae) in Australia. *J. Appl. Ecol.* **25**, 867–87.

Barnes, D. K. A. (1995) Seasonal and annual growth in erect species of Antarctic bryozoans. *J. Exp. Mar. Biol. Ecol.* **188**, 181–98.

Bavestrello, G., Cattaneo Vietti, R., Cerrano, C. & Sara, M. (1993) Rate of spiculogenesis in *Clathrina cerebrum* (Porifera: Calcispongiae) using tetracycline marking. *J. Mar. Biol. Assoc. UK* **73**, 457–60.

Berman, M. S., McVey, A. L. & Ettershank, G. (1989) Age determination of Antarctic krill using fluorescence and image analysis of size. *Polar Biol.* **9**, 267–72.

Bibby, C. J., Burgess, N. D. & Hill, D. A. (1992) *Bird Census Techniques.* Academic Press, San Diego.

Bray, J. R. & Curtis, C. T. (1957) An ordination of the upland forest communities of southern Wisconsin. *Ecol. Monogr.* **27**, 325–49.

Britaev, T. A. & Belov, V. V. (1993) Age determination in polynoid polychaetes, using growth rings on their jaws. *Zoologicheskii Zhurnal* **72**, 15–21.

Broadbent, L., Doncaster, J., Hull, R. & Watson, M. (1948) Equipment used for trapping and identifying alate aphids. *Proc. R. Ent. Soc. Lond. A* **23**, 57–8.

Buckland, S. T., Anderson, D. R., Burnham, K. P. & Laake, J. L. (1993) *Distance Sampling: estimating abundance of biological populations.* Chapman & Hall, London.

Camin, V., Baker, P., Carey, J., Valenzuela, J. & Arredondo Peter, R. (1991) Biochemical age determination for adult Mediterranean fruit flies (Diptera: Tephritidae). *J. Econ. Entomol.* **84**, 1283–8.

Clark, P. I. & Evans, F. C. (1954) Distance to nearest neighbor as a measure of spatial relationships in populations. *Ecology* **35**, 445–53.

Cole, L. C. (1949) The measurement of interspecific association. *Ecology* **30**, 411–24.

Colwell, R. K. & Coddington, J. A. (1994) Estimating terrestrial biodiversity through extrapolation. *Phil. Trans. R. Soc. Lond. B* **345**, 101–18.

Connell, J. H. (1970) A predator–prey system in the marine intertidal region. I. *Balanus glandula* and several predatory species of *Thais*. *Ecol. Monogr.* **40**, 710–23.

Coon, B. F. & Rinicks, H. B. (1962) Cereal aphid capture in yellow baffle trays. *J. Econ. Entomol.* **55**, 407–8.

Corbet, P. S. (1962a) Age-determination of adult dragonflies (Odonata). In: *Proceeding of the XIth International Congress on Entomology* **3**, 287–9.

Corbet, P. S. (1962b) *A Biology of Dragonflies*. Witherby, London.

Coulson, G. M. & Raines, J. A. (1985) Methods for small-scale surveys of grey kangaroo populations. *Aust. Wildlife Res.* **12**, 119–25.

Crossley, D. A. & Blair, J. M. (1991) A high efficiency, "low technology" Tullgren-type soil extractor for soil microarthropods. *Agric. Ecosys. Environ.* **34**, 187–92.

De Jonge, V. N. & Bouwman, L. A. (1977) A simple density separation technique for quantitative isolation of meiobenthos using colloidal silica Lodox™. *Mar. Biol.* **82**, 379–84.

Digby, P. G. N. & Kempton, R. A. (1987) *Multivariate Analysis of Ecological Communities*. Chapman & Hall, London.

Edwards, C. A. (1991) The assessment of populations of soil-inhabiting invertebrates. *Agric. Ecosys. Environ.* **34**, 145–76.

Ekaratne, S. U. K. & Crisp, D. J. (1982) Tidal micro-growth bands in tertidal gastropod shells, with an evaluation of band-dating techniques. *Proc. R. Soc. Lond. B* **214**, 305–23.

Eleftheriou, A. & Holme, N. A. (1984) Macrofauna techniques. In: *Methods for the Study of Marine Benthos* (eds N. A. Holme & A. D. McIntyre). Blackwell Scientific Publications, Oxford.

Elliot, J. M. & Drake, C. M. (1981) A comparative study of seven grabs used for sampling benthic macroinvertebrates in rivers. *Freshwater Biol.* **11**, 99–120.

Elliott, J. M. & Tullet, P. A. (1978) *A Bibliography of Samplers for Benthic Invertebrates*, vol. 4. Freshwater Biological Association Occasional Publication.

Engeman, R. M., Sugihara, R. T., Pank, L. F. & Dusenberry, W. E. (1994) A comparison of plotless density estimators using Monte Carlo simulation. *Ecology* **75**, 1769–79.

Evans, C. R. & Lockwood, A. P. M. (1994) Population field studies of the Guinea chick lobster (*Panulirus guttatus* Latreille) at Bermuda: abundance, catchability, and behavior. *J. Shellfish Res.* **13**, 393–415.

Evans, F. C. & Freeman, R. B. (1950) On the relationship of some mammal fleas to their hosts. *Ann. Ent. Soc. Am.* **43**, 320–33.

Evans, L. J. (1975) An improved aspirator (pooter) for collecting small insects. *Proc. Br. Entomol. Nat. Hist. Soc.* **8**, 8–11.

Feldhamer, G. A. & Maycroft, K. A. (1992) Unequal capture response of sympatric golden mice and white-footed mice. *Am. Midl. Nat.* **128**, 407–10.

Fore, L. S., Karr, J. R. & Wisseman, R.W. (1996) Assessing invertebrate responses to human activities: evaluating alternative approaches. *J. North Am. Benthol. Soc.* **15**, 212–31.

Fowler, J. & Cohen, L. (1990) *Practical Statistics for Field Biology*. Open University Press, Buckingham, UK.

Francillon Viellot, H., Arntzen, J.W. & Geraudie, J. (1990) Age, growth and longevity of sympatric *Triturus cristatus*, *Triturus marmoratus* and their hybrids (Amphibia, Urodela): a skeletochronological comparison. *J. Herpetol.* **24**, 13–22.

Frost, S., Huni, A. & Kershaw, W. E. (1971) Evaluation of a kicking technique for sampling stream bottom fauna. *Can. J. Zool.* **49**, 167–73.

Galinou Mitsoudi, S. & Sinis, A. I. (1995) Age and growth of *Lithophaga lithophaga* (Linnaeus, 1758) (Bivalvia: Mytilidae), based on annual growth lines in the shell. *J. Mollusc. Stud.* **61**, 435–53.

Garvey, J. E., Marschall, E. A. & Wright, R. A. (1998) From star charts to stoneflies: detecting relationships in continuous bivariate data. *Ecology* **79**, 442–7.

Gates, C. E. (1969) Simulation study of estimators for the line transect sampling method. *Biometrics* **25**, 317–28.

Gates, C. E., Marshall, W. H. & Olson, D. P. (1968) Line transect method of estimating grouse population densities. *Biometrics* **24**, 135–45.

Gerard, D., Bauchau, V. & Smets, S. (1994) Reduced trappability in wild mice, *Mus musculus domesticus*, heterozygous for Robertsonian translocations. *Anim. Behav.* **47**, 877–83.

Glasgow, J. P. & Duffy, B. J. (1961) Traps in field studies of *Glossina pallidipes* Austen. *Bull. Entomol. Res.* **52**, 795–814.

Goldsmith, F. B. & Harrison, C. M. (1976) Description and analysis of vegetation. In: *Methods in Plant Ecology* (ed. S. B. Chapman). Blackwells, Oxford.

Goodyear, C. P. (1997) Fish age determined from length: an evaluation of three methods using simulated red snapper data. *Fish. Bull.* **95**, 39–46.

Greenwood, J. J. D. (1996) Basic techniques. In: *Ecological Census Techniques* (ed. W. J. Sutherland). Cambridge University Press, Cambridge.

Gregory, R. S. & Powles, P. M. (1985) Chronology, distribution, and sizes of larval fish sampled by light traps in macrophytic Chemung Lake (Ontario, Canada). *Can. J. Zool.* **63**, 2569–77.

Greig-Smith, P. (1978) *Quantitative Plant Ecology*. Butterworths, London.

Gressitt, J. L. & Gressitt, M. K. (1962) An improved Malaise trap. *Pacific Insects* **4**, 87–90.

Harper, A. M. & Story, T. P. (1962) Reliability of trapping in determining the emergence period and sex ratio of the sugar-beet root maggot *Tetanops myopaeformis* (Rbder) (Diptera: Otitidae). *Can. Entomol.* **94**, 268–71.

Harrison, S., Ross, S.J. & Lawton, J.H. (1992) Beta-diversity on geographic gradients. *J. Anim. Ecol.* **61**, 151–8.

Hayne, D.W. (1949) Two methods of estimating populations from trapping records. *J. Mammal.* **30**, 399–411.

Henderson, P. A. (1989) On the structure of the inshore fish community of England and Wales. *J. Mar. Biol. Assoc UK* **69**, 145–63.

Henderson, P. A. & Holmes, R. H. A. (1991) On the population dynamics of dab, sole and flounder within Bridgwater Bay in the lower Severn Estuary, England. *Netherlands J. Sea Res.* **27**, 337–44.

Henderson, P. A. & Walker, I. (1986) On the leaf-litter community of the Amazonian blackwater stream Tarumazinho. *J. Trop. Ecol.* **2**, 1–17.

Henderson, P. A., Irving, P.W. & Magurran, A. E. (1997) Fish pheromones and evolutionary enigmas: a reply to Smith. *Proc. R. Soc. Lond. B.* **264**, 451–3.

Hill, M. O. (1973) Diversity and evenness: a unifying notation and its consequences. *Ecology* **54**, 427–32.

Hill, M. O., Bunce, R. G. H. & Shaw, M.W. (1975) Indicator species analysis, a divisive polythetic method of classification and its application to a survey of native pinewoods in Scotland. *J. Ecol.* **63**, 597–613.

Hodson, A. C. & Brooks, M. A. (1956) The frass of certain defoliators of forest trees in the north central United States and Canada. *Can Entomol.* **88**, 62–8.

Hunter, W. & Simpson, A. E. (1976) A benthic grab designed for easy operation and durability. *J. Mar. Biol. Assoc. UK* **56**, 951–7.

Hurlbert, S. H. (1969) A coefficient of interspecific association. *Ecology* **50**, 1–9.

Hurlbert, S. H. (1971) The non-concept of species diversity: a critique and alternative parameters. *Ecology* **52**, 577–86.

Jolly, G. M. (1965) Explicit estimates from capture–recapture data with both death and immigration – stochastic model. *Biometrika* **52**, 225–47.

Jongman, R. H. G., Ter Braak, C. J. F. & Van Tongeren, O. F. R. (eds) (1995) *Data Analysis in Community and Landscape Ecology*. Cambridge University Press, Cambridge, UK.

Karlsson, M., Bohlin, T. & Stenson, J. (1976) Core sampling and flotation: two methods to reduce costs of a chironomid population study. *Oikos* **27**, 336–8.

Karr, J. R. (1991) Biological integrity: a long-neglected aspect of water resource management. *Ecol. Applications* **1**, 66–84.

Karr, J. R. & Chu, E. W. (1997) *Biological Monitoring and Assessment: using multimetric indexes effectively*. EPA 235-R97-001. University of Washington, Seattle, WA.

Karr, J. R. & Dudley, D. R. (1981) Ecological perspective on water quality goals. *Environ. Management* **5**, 55–68.

Kempson, D., Lloyd, M. & Ghelardi, R. (1963) A new extractor for woodland litter. *Pedobiologia* **3**, 1–21.

Kent, M. & Coker, P. (1992) *Vegetation Description and Analysis*. John Wiley, Chichester, UK.

Kershaw, K. A. (1973) *Quantitative and Dynamic Plant Ecology*. Griffin, London.

Kethley, J. (1991) A procedure for extraction of microarthropods from bulk soil samples with emphasis on inactive stages. *Agric. Ecosys. Environ.* **34**, 193–200.

Keuls, M., Over, H. I. & De Wit, C.T. (1963) The distance method for estimating densities. *Statist. Neerland.* **17**, 71–91.

Kovner, J. L. & Patil, S. A. (1974) Properties of estimators of wildlife population density for the line transect method. *Biometrics* **30**, 225–30.

Krebs, C. J., Singleton, G. R. & Kenney, A. J. (1994) Six reasons why feral house mouse populations might have low recapture rates. *Wildlife Res.* **21**, 559–67.

Laird, M. L. & Stott, B. (1978) Marking and tagging. In: *Methods for Assessment of Fish Production in Fresh Waters* (ed. T. Bagenal). Blackwell Scientific Publications, Oxford.

Lamb, R. I. & Wellington, W. G. (1974) Techniques for studying the behavior and ecology of the European earwig, *Forficula auricularia* (Dermaptera: Forficulidae). *Can. Entomol.* **106**, 881–8.

Laughlin, R. (1965) Capacity for increase: a useful population statistic. *J. Anim. Ecol.* **349**, 77–91.

Laughlin, R. (1976) Counting the flowers in the forest: combining two population estimates. *Aust. J. Ecol.* **1**, 97–101.

Legendre, P. & Legendre, L. (1998) *Numerical Ecology*, 2nd edn. *Developments in Environmental Modelling*, 20. Elsevier, Amsterdam.

Lincoln, F. C. (1930) Calculating waterfowl abundance on the basis of banding returns. *USDA Circ.* **118**, 1–4.

Lisitsyn, A. P. & Udintsev, G. B. (1955) New model dredges. *Trudy vses. gidrobiol. Obsch.* **6**, 217–22 [In Russian].

Lowman, M. D. (1987) Relationships between leaf growth and holes caused by herbivores. *Aust. J. Ecol.* **12**, 189–91.

Luczak, J. & Wierzbowska, T. (1959) Analysis of likelihood in relation to the length of a series in the sweep method. *Bull. Acad. Pol. Sci. Ser. Sci. Biol.* **7**, 313–18.

McIntosh, R. P. (1967) An index of diversity and the relation of certain concepts to diversity. *Ecology* **48**, 1115–26.

MacLeod, J. (1958) The estimation of numbers of mobile insects from low-incidence recapture data. *Trans. R. Ent. Soc. Lond.* **110**, 363–92.

Magurran, A. E. (1988) *Ecological Diversity and its Measurement.* Chapman & Hall, London.

Magurran, A. E., Irving, P. W. & Henderson, P. A. (1996) Why is fish alarm "pheromone" ineffective in the wild? *Proc. R. Soc. Lond. B* **263**, 1551–6.

Manly, B. F. J. (1990) *Stage-structured Populations, Sampling, Analysis and Simulation.* Chapman & Hall, London.

Martin, F. R., McCreadie, J. W. & Colbo, M. H. (1994) Effect of trap site, time of day, and meteorological factors on abundance of host-seeking mammalophilic black flies (Diptera: Simulidae). *Can. Entomol.* **126**, 283–9.

Mathis, A. & Smith, R. J. F. (1992) Avoidance of areas marked with a chemical alarm substance by fathead minnows (*Pimephales promelas*) in a natural habitat. *Can. J. Zool.* **70**, 1473–6.

Milbrink, G. & Wiederholm, T. (1973) Sampling efficiency of four types of mud bottom sampler. *Oikos* **24**, 479–82.

Mochizuki, A., Shiga, M. & Imura, O. (1993) Pteridine accumulation for age determination in the melon fly, *Bactrocera (Zeugodacus) cucurbitae* (Coquillett) (Diptera: Tephritidae). *Appl. Ent. Zool.* **28**, 584–6.

Mooty, J. J. & Karns, P. D. (1984) The relationship between white-tailed deer track counts and pellet-group surveys. *J. Wildlife Management* **48**, 275–9.

Moran, P. A. P. (1951) A mathematical theory of animal trapping. *Biometrica* **38**, 307–11.

Morris, R. F. (1942) The use of frass in the identification of forest insect damage. *Can. Entomol.* **74**, 164–7.

Murphy, W. L. (1985) Procedure for the removal of insect specimens from sticky-trap material. *Ann. Ent. Soc. Am.* **78**, 881.

Myllymaki, A., Paasikallio, A., Pankakoski, E. & Kanervo, V. (1971) Removal experiments on small quadrates as a means of rapid assessment of abundance of small mammals. *Ann. Zool. Fennici* **8**, 177–85.

Naugle, D. E., Jenks, J. A. & Kernohan, B. J. (1996) Use of thermal infrared sensing to estimate density of white-tailed deer. *Wildlife Soc. Bull.* **24**, 37–43.

Nichols, D. (1983) *Safety in Biological Fieldwork, guidance notes for codes of practice.* Institute of Biology, London.

Nuckols, M. S. & Connor, E. F. (1995) Do trees in urban or ornamental plantings receive more damage by insects than trees in natural forests. *Ecol. Entomol.* **20**, 253–60.

O'Connor, F. B. (1957) An ecological study of the Enchytraeid worm population of a coniferous forest soil. *Oikos* **8**, 162–99.

Odum, H. T., Cantfon, J. E. & Kornicker, L. S. (1960) An organizational hierarchy postulate for the interpretation of species-individual distributions, species entropy, ecosystem evolution and the meaning of a species-variety index. *Ecology* **41**, 395–9.

Otis, D. L., Burnham, K. P., White, G. C. & Anderson, D. R. (1978) Statistical inference from capture data on closed animal populations. *Wildlife Monogr.* **62**, 1–135.

Paramonov, A. (1959) A possible method of estimating larval numbers in tree crowns. *Ent. Man. Mag.* **95**, 82–3.

Parker, K. R. (1979) Density estimation by variable area transect. *J. Wildlife Management* **43**, 484–92.

Patil, G. P. & Taillie, C. (1979) An overview of diversity. In: *Ecological Diversity in Theory and Practice* (eds J. F. Grassle, G. P. Patil, W. Smith & C. Taillie). International Cooperative Publishing House, Fairland, MD.

Pennak, R. W. (1962) Quantitative zooplankton sampling in littoral vegetation areas. *Limnol. Oceanogr.* **7**, 487–9.

Pielou, D. P. & Pielou, E. C. (1968) Association among species of infrequent occurrence: the insect and spider fauna of *Polyporus betulinus* (Builiard) Fries. *J. Theoret. Biol.* **21**, 202–16.

Powers, C. F. & Robertson, A. (1967) Design and evaluation of an all-purpose benthos sampler. *Great Lakes Research Division, Spec. Rep.* **30**, 126–31.

Putman, R. J. (1995) Ethical considerations and animal welfare in ecological field studies. *Biodivers. Conserv.* **4**, 903–15.

Raboud, C. (1986) Age determination of *Arianta arbustorum* (L.) (Pulmonata) based on growth breaks and inner layers. *J. Mollusc. Stud.* **52**, 243–7.

Redfern, M. (1968) The natural history of spear thistle-heads. *Field Stud.* **2**, 669–717.

Renyi, A. (1961) On measures of entropy and information. In: *Proceedings of the 4th Berkely Symposium on Mathematical Statistics and Probability* (ed. J. Neyman), pp. 547–561. University of California Press, Berkley, CA.

Ricker, W. E. (1975) *Computation and Interpretation of Biological Statistics of Fish Populations.* Bulletin of the Fisheries Board, Ottawa.

Robinson, H. S. & Robinson, P. J. M. (1950) Some notes on the observed behavior of Lepidoptera in flight in the vicinity of light-sources together with a description of a light-trap designed to take entomological samples. *Entomol. Gaz.* **1**, 3–15.

Routledge, R. D. (1977) On Whittaker's components of diversity. *Ecology* **58**, 1120–7.

Seber, G. A. F. (1982) *The Estimation of Animal Abundance and Related Parameters.* Griffin, London.

Sheehy, M. R. J., Greenwood, J. G. & Fielder, D. R. (1994) More accurate chronological age determination of crustaceans from field situations using the physiological age marker, lipofuscin. *Mar. Biol.* **121**, 237–45.

Sheehy, M. R. J., Shelton, P. M. J., Wickins, J. F., Belchier, M. & Gaten, E. (1996) Ageing the European lobster *Homarus gammarus* by the lipofuscin in its eyestalk ganglia. *Mar. Ecol. Progr. Ser.* **146**(1–3), 99–111.

Simpson, E. H. (1949) Measurement of diversity. *Nature* **163**, 688.

Skidmore, P. (1985) *The Insects of the Cowdung Community.* AIDGAP guide produced by the Field Studies Council. The Dorset Press, UK.

Slobodkin, L. B. (1962) *Growth and Regulation of Animal Populations.* Holt, Rinehart and Winston, New York.

Smith, B. (1986) *Evaluation of Different Similarity Indices Applied to Data from the Rothamsted Insect Survey.* University of York, York.

Smith W. & McIntryre, A. D. (1954) A spring-loaded bottom sampler. *J. Mar. Biol. Assoc. UK* **33**, 257–64.

Sokal, R. R. & Rohlf, F. J. (1995) *Biometry: the principles and practice of statistics in biological research*. W. H. Freeman, New York.

Solomon, J. D. (1977) Frass characteristics for identifying insect borers (Lepidoptera: Cossidae and Sesiidae; Coleoptera: Cerambyciidae) in living hardwoods. *Can. Entomol.* **109**, 295–303.

Southwood, T. R. E. (1996) Natural communities: structure and dynamics. *Phil. Trans. R. Soc. Lond. B.* **351**, 1113–29.

Southwood, T. R. E. & Henderson, P. A. (2000) *Ecological Methods*, 3rd edn. Blackwell Science, Oxford.

Southwood, T. R. E., Brown, V. K. & Reader, P. M. (1979) The relationship between insect and plant diversities in succession. *Biol. J. Linn. Soc.* **12**, 327–48.

Stewart, A. J. A. & Wright, A. F. (1995) A new inexpensive suction apparatus for sampling arthropods in grassland. *Ecol. Entomol.* **20**, 98–102.

Suckling, G. C. (1978) A hair sampling tube for the detection of small mammals in trees. *Aust. Wildlife Res.* **5**, 249–52.

Sunderland, K. D., Hassell, M. & Sutton, S. L. (1976) The population dynamics of *Philoscia muscorum* (Crustacea, Oniscoidea) in a dune grassland ecosystem. *J. Anim. Ecol.* **45**, 487–506.

Thomas, D. B. & Chen, A. C. (1989) Age determination in the adult screwworm (Diptera: Calliphoridae) by pteridine levels. *J. Econ. Entomol.* **82**, 1140–4.

Thomas, D. B. & Chen, A. C. (1990) Age distribution of adult female screwworms (Diptera: Calliphoridae) captured on sentinal animals in the coastal lowlands of Guatemala. *J. Econ. Entomol.* **83**, 1422–9.

Thomas, J. A. (1983) A quick method for estimating butterfly numbers during surveys. *Biol. Conservation* **27**, 195–211.

Tinbergen, L. (1960) The natural control of insects in pinewoods. 1. Factors influencing the intersity of predation by song birds. *Arch Neérl. Zool.* **13**, 266–343.

Topping, C. J. & Sunderland, K. D. (1992) Limitations to the use of pitfall traps in ecological studies exemplified by a study of spiders in a field of winter wheat. *J. Appl. Ecol.* **29**, 485–91.

Tóthméresz, T. (1995) Comparison of different methods for diversity ordering. *J. Vegetation Sci.* **6**, 283–90.

Townes, H. (1962) Design for a Malaise trap. *Proc. Entomol. Soc. Wash.* **64**, 253–62.

Turner, D. C. (1975) *The Vampire Bat*. Johns Hopkins Press, Baltimore, MD.

Underwood, A. J. (1997) *Experiments in Ecology: their logical design and interpretation using analysis of variance*. Cambridge University Press, Cambridge.

Van Veen, J. (1933) Onderzoek naar het zandtransport von rivieren. *De Ingenieur* **48**, 151–9.

Waite, S. (2000) *Statistical Ecology in Practice*. Prentice Hall, Harlow, UK.

Walsh, P. M., Halley, D. J., Harris, M. P., del Nevo, A., Sim, I. M. & Tasker, M. L. (1995) *Seabird Monitoring Handbook for Britain and Ireland*. JNCC, Peterborough.

Walter, D. E., Kethley, J. & Moore, J. C. (1987) A heptane flotation method for recovering microarthropods from semiarid soils, with comparison to the Merchant–Crossley high-gradient extraction method and estimates of microarthropod biomass. *Pediobiologia* **30**, 221–32.

Walther, B. A. & Morand, S. (1998) Comparative performance of species richness estimation methods. *Parasitology* **116**, 395–405.

Wastle, R. J., Babaluk, J. A. & Decterow, G. M. (1994) A bibliography of marking fishes with tetracyclines including references to effects on fishes. *Can. Tech. Rep. Fish. Aqu. Sci.* **1951**, 1–26.

Webster, J. P., Brunton, C. F. A. & MacDonald, D. W. (1994) Effect of *Toxoplasma gondii* upon neophobic behaviour in wild brown rats, *Rattus norvegicus*. *Parasitology* **109**, 37–43.

Weiss, H. B. & Boyd, W. M. (1950) Insect feculae I. *J. NY Entomol. Soc.* **58**, 154–68.

Weiss, H. B. & Boyd, W. M. (1952) Insect feculae II. *J. NY Entomol. Soc.* **60**, 25–30.

White, G. C. & Eberhardt, L. E. (1980) Statistical analysis of deer and elk pellet group data. *J. Wildlife Management* **44**, 121–31.

Whittaker, R. H. (1952) A study of summer foliage insect communities in Great Smoky Mountains. *Ecol. Monogr.* **22**, 1–44.

Whittaker, R. H. (1972) Evolution and measurement of species diversity. *Taxon* **21**, 213–51.

Whittaker, R. H. & Fairbanks, C. W. (1958) A study of plankton copepod communities in the Columbia basin, south eastern Washington. *Ecology* **39**, 46–65.

Wilson, M. V. & Mohler, C. L. (1983) Measuring compositional change along gradients. *Vegetatio* **54**, 129–41.

Wilson, M. V. & Schmida, A. (1984) Measuring beta diversity with presence–absence data. *J. Ecol.* **72**, 1055–64.

Wolda, H. (1981) Similarity indices, sample size and diversity. *Oecologia* **50**, 296–302.

Wolda, H. (1983) Diversity, diversity indices and tropical cockroaches. *Oecologia* **58**, 290–8.

Woolhouse, M. E. J. & Chandiwana, S. K. (1990) Population biology of the freshwater snail *Bulinus globosus* in the Zimbabwe highveld. *J. Appl. Ecol.* **27**, 41–59.

Zar, J. H. (1996) *Introductory Biostatistics.* Prentice Hall, New York.

Zippin, C. (1956) An evaluation of the removal method of estimating animal populations. *Biometrics* **12**, 163–89.

Zippin, C. (1958) The removal method of population estimation. *J. Wildlife Management* **22**, 82–90.

Index